Carole Acaturky

LIFE WITH A CHANNEL SURFER

Also by Carole Achterhof

NEVER TRUST A SIZE THREE
(ISBN 0-9625940-0-8)

POTATO CHIPS ARE VEGETABLES
(ISBN 0-9625940-1-6)

LIFE WITH A CHANNEL SURFER

Written and Illustrated by

CAROLE ACHTERHOF

Cover design by KEITH VAN WESTEN

Bare Bones Books
Luverne, MN 56156

Dedicated to my students at
Adrian (MN) High School

Bare Bones Books
315 North Freeman
Luverne, MN 56156

Library of Congress Catalog Card number: 94-70345
ISBN 0-9625940-2-4

Printed in the United States by
Crescent Publishing, Inc., Hills, MN 56138

First Printing

TABLE OF CONTENTS

Life With a Channel Surfer

Although we've never met, Bob Thomas is my hero. The Spokane, Wash., engineer ranks right up there with the inventors of microwave ovens, elastic waistbands and disposable diapers.

Thomas has designed a gadget to disarm the remote controls of channel-surfers and he has won my heart. Called "Stop It!", the device momentarily blocks the remote control's infrared signals, making channel changes impossible.

According to an article in the Wall Street Journal, Thomas thinks his new invention will "make a man understand how frustrating it can be to have the channels changed on you."

When it comes to channel-surfers, I speak from experience. Actually, the surfer in our house is more like a channel jet-skier. In the time that it takes me to groan, get up from the couch and leave the room, he can flip through 33 channels without blinking an eye. When it comes to the remote control, he is the fastest draw this side of the OK Corral.

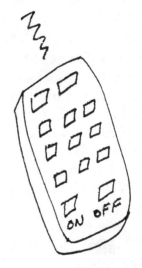

In the same article, Howard Markman, head of the Center for Marital Studies at the University of Denver, noted that men need to hold a remote control as a "vestige of holding a weapon."

As women have suspected all along, the remote control is part of the power thing.

Markman also suggested that women should

"snuggle up to the channel-surfer and have them hold on to you instead of the remote control."

His solution may be easier said than done. For starters, I seriously doubt whether my husband's reclining chair could hold two persons and a basket of clothes to be folded, my usual evening entertainment.

Click. Click. Click.

"Do you mind if I climb into that chair with you? Don't bother with moving over. An article I just read said that we should snuggle more often."

Click. Click. Click. "What did you say?"

"Never mind. Just pretend I'm not here. There. Isn't this cozy? Would you mind terribly reaching for that basket behind me? My right arm seems to be pinned against the side of the chair."

Click. Click. Click.

"This is great. Just great. Cuddled up next to the man I love and folding a big basketful of nice, clean socks. Life doesn't get much better than this. Actually, folding socks with one hand isn't that difficult once you get used to it."

Click. Click. Click.

"I really hate to bother you, but I was wondering. If people lose all feeling in their legs and their toes start turning purple, should they be concerned?"

Click. Click. Click. "Mmmm. Did you say something?"

Click. Click. Click

"No. No. Now that I seem to be paralyzed from the neck down, my legs don't seem to matter anymore. Purple toes are good."

Click. Click. Click.

"Look, this isn't going to work. Could I ask you one more question?"

Click. Click. Click. "Uh, uh. What is it?"

"Do you still have that crowbar?"

8

There's no doubt about it. Investing in a $29.95 "Stop It!", would be much less hazardous to my health than snuggling.

Remotely Speaking

TV remote controls were a bad idea.

Ranked with other domestic issues, such as the fears of a nuclear attack or running out of toilet paper, having no control over a television set is the greatest cause of troubled relationships in our country.

While one person in every household flaunts his or her power and flips non-stop through hundreds of television programs, the rest of us are stomping off to other rooms. No matter how much we sigh and complain about never seeing more than 10 minutes of any given program, our laments fall on deaf ears.

Harper Barnes, a writer for the St. Louis Post-Dispatch, recently offered his reason for this phenomenon. According to Barnes, half an hour in front of a flickering TV screen will turn anyone into a zombie and knock 50 points off his or her IQ.

Although most of my knowledge about zombies comes from reading "The Night of the Living Dead", I'm fairly certain that interpersonal relationships are the least of their worries. There's also a very good possibility that a person sitting down to watch TV for a couple of hours after supper will be on par intellectually with the tossed salad he's just consumed.

A scientist in the 1978 book, "Four Arguments for the Elimination of Television", noted, "Once the set goes on, the brain waves slow down...the longer the set is on, the slower the brainwave activity."

It's all because televisions send out 30 images a second and most viewers, unable to compute 30 of anything per second, simply short-circuit. Very possibly, they are unaware that we've stomped out of the room.

As if living with a remote control power-monger wasn't enough to make anyone question old wedding vows or start a search for one's real parents, the situation has become more complicated. A new TV on the market allows viewers to watch four channels at the same time. Collectively, that's 120 images a second and we're no longer talking about short-circuiting. We're talking melt-down.

Compounded with the capacity to flip through scores of TV channels each minute with a remote control, this new TV viewing may mean the end of civilization as we once knew it.

It's all too much too fast. It seems like only yesterday when I was young and my family received only one channel on their 10-inch screen. Even though early programming filled a precious two hours a day, we would often sit transfixed in front of the set, watching the test pattern.

As children, we would invite our friends over to watch the test pattern. For hours, we would stare at the black-and-white profile of the American Indian in full headdress, surrounded by tiny patterns of horizontal and vertical lines. It was truly a wonder to behold.

Best of all, we only had to deal with one image every 22 hours. With all of our faculties intact, we would have noticed someone stomping out of the room.

Goose Livers and TV

Englishman Sydney Smith (1771-1845) was a contradiction - he was both a clergyman and a humorist. Although it's difficult to imagine a combination of Billy Graham and Richard Pryor, Smith did come up with some interesting observations.

He once wrote, "My idea of heaven is eating foie gras to the sound of trumpets."

My husband would most likely pass on the goose livers. Like many husbands, he would settle for a heaven equipped with a reclining chair, a thousand cable stations and a remote control.

There was once a time when we were both purists about television, particularly when we first began going to our lake cabin. As we enjoyed the good life - chopping wood for our fires, boiling water for the babies and cleaning fish - we didn't do television.

We laughed at the nearby state park with its cable TV hookups.

The purist life-style went well - for about a week. A compromise was soon reached and we purchased a small, basic, black-and-white television. Of course, it would be watched only during menacing weather and threats to national security.

Because there was only one channel to watch, life continued to be simple. I screamed at spiders and he would kill them.

By the second week, I began wondering if "Guiding Light's" Lisa had found a fourth husband and whether Oprah Winfrey had lost any more weight. I missed weather maps and commercial jingles. As we suffered the pangs of video withdrawal, the reasons to turn on the set became more frequent - we had to check the horizontal hold, the antenna connection and whether the lightning storm had affected the set.

Three days of non-stop rain finally put us over the edge. We were hooked. We also questioned whether hand-chopped wood burned any better than wood purchased from a nearby neighbor.

Unlike love, our old set didn't last forever. It was replaced last week with a 1992 model, complete with automatic tuning, sleep alarm and a remote control. High technology had entered our lives.

Unfortunately, we still only received one channel. Words can't describe the poignancy of watching a man with a remote control in his hand and one available station. Quite automatically, my husband switched from Channel 12 to Channel 12 to Channel 12.

As the same program kept reappearing on the screen, I could better understand his alternative to heaven. Even goose livers and blaring trumpets would be better than having only one TV channel.

Too Tired to Exercise

The definition of physical exercise is rapidly changing. If the current trend continues at its present rate, I may soon be thought of as "athletic".

The latest government guidelines say that 30 minutes of moderate, intermittent exercise - such as a brisk walk, stair climbing or gardening - do almost as much good as one rigorous workout in a gym.

That's good news for those of us who would rather pick tomatoes than figure out the difference between low impact and high impact aerobics. When it comes to facing the harsher realities of life, I would much rather deal with rot spots on cucumbers than how to squeeze my body into a couple ounces of Spandex.

In light of the fact that I have been surrounded by physical fitness buffs all summer, the report is definitely a relief. I've been playing hostess to a steady stream of visitors who choose to fill their days with jogging, walking, swimming, wind surfing and bicycling.

All of their self-inflicted tortures give new meaning to the expression, "Different strokes for different folks."

"Let's go for a nice, long walk", urged one of them last week. According to her reasoning, General Sherman's march from Atlanta to the sea was also a nice, long walk.

Instead, I stayed home. I washed five loads of laundry, ran up and down the basement stairs at least 20 times, prepared two complete meals and stored them in the refrigerator, and changed the sheets on five beds.

When my visitor returned from her run, she said, "After a shower, I think I'll lie down for a few minutes. I'm exhausted."

While she slept, I vacuumed the floors, dusted every surface in sight and folded a mountain of bath towels.

After lunch, I loaded the dishwasher and washed the car, and she announced, "What I need to do is ride the stationary bike. I really need the exercise."

Feeling somewhat guilty about my sedentary lifestyle, I retreated to the garden, where I pulled up delinquent weeds and thinned out the carrots.

When I walked back inside, she was dozing off again. I tried not to wake anyone as I set the table and finished fixing the dinner. As I glanced at the picture of health draped over the armchair, I promised myself to begin exercising the next day. She looked so healthy. So well-rested.

While I sluggishly held back and stayed at home after dinner, she went running for the longest time. In fact, by the time she returned, I had cleaned up the kitchen and scrubbed all the surfaces in the bathroom.

I also washed another load of towels. People who exercise use a lot of towels.

Before she turned in for the night, she advised, "You really should go running with me in the morning. There's nothing like it."

Long after she had gone to bed, and I had watered the house plants and picked up the living room, I was still feeling guilty about not getting enough exercise.

I guess I would like to exercise, but I'm too tired.

Lessons From a Fashion Doll

The library has an unusual collection of fashion dolls on display. What makes the display so unusual is that the dolls are still in their original boxes.

As I looked at the untouched figures of Barbie, Ken, Midge, Alan and little Skipper, I realized I had never seen a Barbie doll in its original container for more than six seconds. By the time one of our daughters tore off a package's ribbon and paper, the doll had already been undressed and was in its second outfit.

I also thought about the lessons learned by living with a Barbie doll that may be carried through life:

Appearance is everything. Life is little more than getting dressed up and then changing clothes again.

There are bad haircuts. Whether your hair has been butchered by an eight-year-old wielding blunt scissors and fingernail clippers or an adult beautician with a bad attitude, you may end up looking like a well-worn scrub brush.

If your legs are bent the wrong way enough times, they will eventually fall off.

Even if they are reattached, the arms and legs will never work the same.

It's impossible to apply lipstick to one-fourth inch lips with a full-sized tube.

It's rare to find shoes that fit. If you find such a pair, they won't last long. They'll either be vacuumed up or you will lose them.

If you meet a boy whose hair looks like plastic, it probably is.

Not all men are anatomically correct.

Just because your younger brother rips off your Barbie's head, it's no proof he will grow up to be a criminal pyschopath. If he hangs Barbie from a second story window with your father's shoestring, there's a slight chance he won't spend most of his life making license plates.

It's impossible to pull a tight sleeve up your arm if your fingers are outstretched.

No matter how nice they look when you buy them, button-up blouses leave gaps in the most embarrassing places.

If you buy a pink convertible, the wheels will fall off within two weeks.

It's difficult to sit in a car if your knees don't bend.

It's important to have big feet. If your measurements in inches are 4-2-3 and you have half-inch feet, you will fall on your face.

Always keep smiling - even if someone puts you in a toilet bowl and pretends it's a swimming pool.

Names are important. A girl named Skipper will never be taken seriously. It's almost a certainty she will never be named to the Supreme Court.

If someone draws on your skin with indelible ink, you're marked for life.

All brides are beautiful. Even if you're covered with crayon and ink marks and your hair looks like a scrub brush, a wedding dress will work wonders.

Toilet Seat Blues

Until the builders arrived at our house, I never imagined having problems with a toilet seat.

I also discovered that you're never too old to develop a fear of falling. Even if you're in a familiar room and the door is locked, the fear can be there.

My first experience with free-falling into a toilet bowl happened after work began on a new addition to the house and I had retreated to the reading room.

The gut-wrenching, sudden descent caught me completely off guard. For five decades I have had built-in instincts about how far the seat of the commode is from the bathroom floor.

After all, bathroom fixtures are permanent objects. Like walls and floors, they're not expected to change positions from one moment to the next.

My first thought, as I felt myself falling through space, was that I had suddenly grown taller and had missed the mark. After my life had flashed before my eyes and I had been immersed in two gallons of ice cold water with my feet pointed at the ceiling, I reached a more logical conclusion.

A man had left the seat up.

I could suddenly understand why women in coed dorms are filing class action suits and why other women are taking their spouses to marriage counselors. Long-suffering women are finally breaking their code of silence for the first time since Thomas Crapper became the father of the flush toilet.

The great, unsolved mystery for me this summer is why certain humans, able to build entire houses with their bare hands and capable of operating heavy duty power equipment, are unable to lower a one-pound seat.

Like many other people, I was taught to leave a room the way I found it. In the same way that we close a refrigerator or oven door after it's opened, toilet seats were meant to be lowered. Returning a seat to its down position should be as natural as closing the door as we leave the house or closing car doors before we drive.

Actually, that leads us to a fairly safe assumption. Anyone who would drive with his car door open would most likely leave a toilet seat up.

As a result of my tumble into the toilet, I will never again walk into a bathroom assuming all is right with the world.

I will also remember the words of Theodore Roosevelt, who once advised, "Keep your eyes on the stars and your feet on the ground."

The Bait Shop

For a few, horror-filled minutes this summer, my refrigerator resembled a bait shop. Even though I've grown used to seeing lower life forms on the shelves, leeches are a completely different can of worms.

It's not as though I've never been repulsed by the abandoned world of refrigerated foods. I've watched navel oranges grow lint, lettuce heads drown in their own juices and leftovers turn green after spending a few months in plastic containers.

I've seen a jar of pickle relish tip over on the back of a refrigerator shelf and drench a cherry cheesecake. I've watched in horror as trails of unidentifiable liquids have worked their way to the vegetable crisper.

It was no coincidence that the leeches appeared when our nephew, an avid fisherman, moved in for the summer.

Shortly after his arrival, I opened the refrigerator door and discovered several styrofoam containers perched precariously on the shelves. Hoping beyond hope that our nephew had done the impossible and found a Chinese carry-out restaurant, I seized the nearest container.

Instead of finding Moo Goo Gai Pan, I found myself eyeball-to-eyeball with a black, slimy leech. On the other hand, it might have been eyeball-to-sucker. It's almost impossible to establish firm eye contact with a leech.

Much to my dismay, the other containers held dirt-crusted earthworms and more leeches.

When man was given dominion over animals, I

don't think there was any mention of having to entertain them in one's home or providing room and board for them in a refrigerator. We shouldn't have to see leeches or earthworms every time we have a craving for a nighttime snack. It's pretty hard to get excited about eating anything that has been keeping regular company with a cup of bait.

Food for human consumption and fishing bait simply don't mix. I'm also convinced that the first artificial fishing lure was invented by a woman with similar hang-ups about live bait.

As quickly as they had appeared, the slimy creatures disappeared. Our nephew, who's really in touch with other people's feelings, sensed something was wrong when I began screaming hysterically, stamping my feet and slamming cupboard doors.

"This is a kitchen for humans, not a bait shop!" I finally sputtered. "Either the worms and leeches go or I go."

Although the slimy little creatures of this world may beg to differ, he made a wise choice.

How to Raise Your I.Q.

Classical music was given an unusual endorsement by the University of California at Levine. A study done by the school suggests that listening to a 10-minute Mozart recording can raise the final scores on intelligent tests.

Although Mozart's influence is short-lived, 15 minutes to be exact, I.Q. scores rose by as many as nine points. The report suggests that classical music enhances abstract thinking.

It also suggests that my mother listened to too many songs by Frank Sinatra and Perry Como. My I.Q. scores from high school resemble my present bowling scores and I'm a terrible bowler.

If the 15-minute enlightenment brought about by Mozart is to be taken seriously, we can only assume that a conversation about splitting the atom could suddenly downgrade into prattle about hairdos and split ends. A conversation about Einstein's theory of relativity could be reduced to the current plot on a favorite soap opera.

Before I began writing this column, I scanned the local radio stations, hoping to be suitably inspired. It should be noted that classical music stations in my corner of the world are as rare as mosquitoes in the winter. The first station I found was playing country western music. After listening to a few songs based on one-sided love and heartless bankers, I felt strangely compelled to buy a pickup truck, a belt with a large buckle and a used quitar. I could suddenly understand how "Lucille" could leave her husband, four hungry children and crops in the field.

An easy-listening station also failed to elevate my powers of abstract thinking. The listening was so easy that my husband had to call my name four times before I responded. It was the closest I've ever come to an out-of-the-body experience.

Only seconds before I began writing, I tuned into a public radio station and caught the final chords of Mozart's Sonata for Two Pianos in D Major. It may have been my imagination, but for a few fleeting seconds I was sure that I could conjugate verbs in any foreign language or outplay any chess champion. The outline of a summit meeting for world peace, complete with  room assignments and coffee breaks, flitted through my mind.

As suddenly as those illuminations appeared, they disappeared. It might have helped to hear the entire sonata. Genius is a fickle, restless thing. Restless like Jack on "Young and the Restless". Now that Victor has returned from his alleged car accident, what will happen to Jack and Nicky? Will Victoria find out she is engaged to her half-brother?

More Mozart, please.

Furniture - You Can Take It With You

Certain mailrooms must have slow days.

Two enterprising mailroom workers in Charlotte, N.C., have invented a new piece of furniture that can be used as a storage box or a television stand. After the owner has literally watched his or her last sitcom, the birchwood stand can easily be converted into a coffin.

It's called a Kas-Kit and five have been sold so far. The prices are far below those of your usual polished teakwood with brass handles and satin-lining models.

No matter what the price or what it's called, I still can't help thinking that it's just another piece of furniture to dust. Considering the fact that most of my table tops haven't seen the light of day since the Carter Administration, buying this particular piece of furniture could prove to be a grave mistake.

On an emotional level, I'd like to think I deserve something better in the end than a television stand, complete with furniture polish buildup, vacuum cleaner scrapes and fingerprints from buttered popcorn.

Although the Kas-Kit might prove to be an interesting conversation piece when entertaining, people might question the propriety of loading it down with chip dips and other party snacks.

If I have to be buried in a storage container from our home, I'd rather choose the refrigerator. After all, it's been my second home for most of my adult life. I've actually tanned by its light and I've spent countless hours staring blankly into its interior.

If the ancient Egyptian rulers could be entombed

with their favorite foods and a couple of unlucky servants, why shouldn't I embark on that final journey with a few generous portions of leftover lasagna?

My container might even lead to extra compliments long after I'm gone.

"You know, she never let us down. No matter how hot things would get, she always stayed cool."

Although my survivors would no doubt be referring to the refrigerator rather than its contents, a compliment is a compliment. Being interred in the refrigerator would also guarantee that I would be missed longer than usual - particularly during television commercials and those interminably long football replays.

Whether the Kas-Kit is used as a television table, a toy box or a recyclable bin, it does manage to disprove one longtime proverb. Contrary to popular belief, you can take it with you.

How to Avoid Housework

It was a case of looking to the heavens for a miracle and not being disappointed.

For several days last fall, I was blissfully freed from domestic chores. There were no floors to vacuum, no tabletops to dust and no toilet bowls to scrub. My liberation from housework happened at the precise moment I heard about a two-ton chunk of metal falling from a misguided Chinese satellite.

Although every scientist predicted a direct hit in the Atlantic Ocean, I had my doubts. After all, if TV weathermen can't accurately predict the weather one day a week, how could anyone trust the scientists? Locating the exact landing spot for the chunk of metal would be no easier than locating the car keys in the bottom of my purse.

In light of having a direct hit on our house, I questioned the value of cleaning the possible target. It was highly unlikely that anyone would comment on a floor clean enough to eat on after it had been reduced to a stack of kindling. Who would criticize the piles of stuff on the upstairs steps when the steps themselves are distributed over a four-block area?

Once I began thinking of our house as "ground zero", I was able to reflect upon the reasons I've had in the past for not doing housework.

In addition to the threat of having objects fall on the house from outer space, it's also foolish to bother cleaning during tornado seasons. In the worst case scenario, it's highly doubtful that a newspaper would run the headline, "Family escapes tornado as immaculate house is demolished."

It's also a good idea to avoid housecleaning on holidays. If you're not satisfied with the routine holidays - Christmas or Valentine's Day - try creating your own. How about the day Hiawatha field-dressed his first deer or the anniversary of your root canal work?

If you really want to take a vacation from housework, cover your front door with yellow plastic tape, giving your home the look of a recent crime scene. Not only will the tape discourage unwelcome "drop-ins", but the ruse could go on indefinitely. If anyone asks, simply explain that you're waiting for the fingerprint experts to arrive.

Another good excuse for not doing housework is having children underfoot. Even if your last baby left home 20 years ago, a cloth diaper thrown over a living room lamp can provide a nice touch. If you're really serious about creating special effects, hire a few neighborhood children to scatter cookie crumbs on the floor and leave sticky smudges on the refrigerator door.

Having a pet is also a good excuse for a not-so-perfect house. If the doorbell rings, throw newspapers over every surface in sight and be ready to explain how much trouble you're having housebreaking "Killer". There's a very good chance that your visitor won't stay long enough to meet your non-existent dog.

If all other excuses fail, there's always the possibility of having another satellite break up in space.

The Toaster Museum

A used toaster museum is the latest attraction in Seattle. Eric Norcross, the museum's founder and curator, is proud of what he thinks is the only museum of its kind. In a newspaper article he referred to toasters as "little chrome monuments to man's ingenuity."

I hate to rain on his parade or dampen his toast, but his museum isn't the first. The first display of unusable and broken toasters, coffee makers and other small appliances had its beginning in our basement over 30 years ago.

While I can vividly recall hauling the useless appliances to the trash can, they have mysteriously reappeared like so many soldiers standing at attention on the basement shelves. Lacking heating coils, prongs on plugs and other essential parts such as legs and lids, they have come to represent my husband's eternal optimism.

He may view them as a parts department for other appliances, but I continue to insist, "If you can't stand or heat, get out of the kitchen."

Actually, ours is a mixed marriage - he's a keeper and I'm a tosser. Once a toaster begins holding bread hostage or a coffee maker begins making strange sounds, I'm ready to pull the plug and buy another. My husband, on the other hand, believes that by adding an electrical cord from a hair dryer or a heating element from an old coffee pot, any appliance will receive a new lease on life.

If there was a religion based on appliance reincarnation, my husband would be the high priest. He will throw away no appliance before its time. While I see appliances as being expandable purchases, he views them as lifelong relationships.

One of the appliances stored away on a basement shelf is a demonic toaster oven, which should have become part of the county landfill years ago. During the first time it was used, it burned a hole in the kitchen counter and I refused to use it again. Call me crazy, but I don't think a craving for a grilled cheese sandwich should result in having a house burn to the ground.

It now sits on a shelf in the dark basement, waiting to be reactivated as replacement parts for coffee makers and hair curlers. One of my greatest fears is someday plugging in an appliance with potentially dangerous toaster oven parts. A person can only explain away so many scorch marks on a kitchen counter.

Knowing that a museum in their city is filled with discarded appliances should make residents of Seattle very uneasy. It also gives new meaning to the title of a recent movie - "Sleepless in Seattle".

Leg Laundries

Members of the Chindogu Society, an inventors' group in Japan, have apparently gone over the edge. Two of their recent inventions manage to combine my least favorite activities - physical exercise and housework.

According to *Details Magazine*, one enterprising soul has invented water-filled compartments that strap on your legs, enabling you to wash clothes while walking vigorously. The second device is a rack worn on your back, secured by a shoulder brace, on which clothing can be hung to dry while you bicycle around town.

Rather than devising ways to eliminate housework, the inventors have gone out of their way to glorify it. The only other undesirable activities still awaiting their attention are having dentist appointments, picking lint balls off sweaters and cleaning bathrooms. Undoubtedly, those activities are on the drawing boards.

If I were to walk downtown with laundry compartments on my legs, the comments of bystanders would be fairly predictable.

"Take a look at her legs!"

"She certainly has her mother's ankles."

"Yes, and her husband's shirts, too!"

If the laundry compartments catch on, complete strangers will be able to tell what kind of wash you're doing. Heavy duty loads would require jogging and spin cycles would only be possible on a dance floor.

"Did you notice how slowly she was walking?"

"Uh, huh. She must have been washing delicates."

As with everything else, leg laundries could become competitive. Typical physical fitness workouts for beginners would include 20 push-ups, 20 sit-ups and two batches of blue-jeans. Serious body builders could boast about washing heavy throw rugs on their more developed legs.

Even romantic strolls would take on an added dimension of realism.

"How about a walk in the moonlight? By the way, don't forget your dirty clothes."

Drying clothes while bicycling would prove even more challenging. Personally, I find it difficult enough to steer, pedal and watch out for traffic at the same time. I simply would be unable to keep track of lost clothespins and trailing bed sheets.

While it would be embarrassing to bicycle through town with my lingerie flying like a cape from my back, dragging a king-size sheet through a downtown intersection could prove to be downright dangerous.

The Japanese inventors apparently never had a grandmother like mine. She often admonished, "Never show your dirty laundry in public."

Sinus Problems

Before moving to Minnesota I knew very little about sinuses. Of course, I had never lived in a place where furnaces blast hot air 10 months of the year, drying out every piece of furniture and nasal passage in sight.

Sinus problems are so common in this part of the country that sufferers are able to complain about their maladies in almost shorthand fashion - "I've got sinuses." Of course they have sinuses, but everyone seems to know what they're talking about.

For every person with sinus problems, there's another person who doesn't have them - and feels guilty about it.

Special shrines - otherwise known as humidifiers - are installed in most homes in an attempt to appease the gods of sinuses. Daily, almost religiously, the coffers of the shrine are filled with water, lime de-scalers and antibacterial agents. They're not exactly human sacrifices, but they seem to do the trick.

The owners of humidifiers seem to be saying, "All right, I've lugged enough gallons of water today to sustain a small country. I've suffered enough. Now let me breathe." It's the sinus sufferer's version of "Let's Make a Deal".

Unfortunately, dumping gallons of water daily into four or five rooms does have drawbacks for those of us who like to breathe without drowning. In our house, for example, the moisture-filled air is not unlike that of a South American rain forest. The plants, which may soon have to be cut down with a machete, love it, but I subconsciously keep an eye out for snakes.

Walking around in a fog isn't all that uncommon, except when you happen to be in your own living room.

Granted, I feel guilty enough about being able to breathe dry air, but there comes a time when any house reaches its saturation level.

Certain guidelines should be established early in a relationship. For instance, the "normal" setting on the humidifier should be more than satisfactory for most human beings. The "high" setting should be reserved for bringing arid deserts back to life.

When books on the shelves expand to twice their normal size and moss is found growing on the walls, I seriously wonder whether I exchanged wedding vows with Aqua Man.

It's not normal to wipe condensation off the television screen before settling down for the evening. Newspapers and magazines shouldn't have to be wrung out before they're read.

As I mentioned before, I didn't know I had sinuses before moving to Minnesota. Because of the humidifier, we've become well acquainted.

Real-Life Dolls

Watching Saturday morning television can be a frightening experience. While I can live with Teenage Mutant Ninja Turtles and the antics of Garfield, one particular commercial manges to send chills up and down my spine.

The "mommy is having a baby doll" commercial features two adorable, little girls fussing over a perky, blonde doll with a frozen smile on her face and a swollen stomach. The accessories for the doll include a baby doll, capable of wetting its diaper, a baby scale and a tiny, square box resembling an ultrasound machine.

As the commercial draws to an end, one of the girls whispers, "Aren't they both pretty?" The implication is that the baby miraculously pops out and that the mother doll is ready for a game of tennis. The most frightening possibility is that, if for some reason, the delivery doesn't go according to plan, the same baby can return to its launch pad and reenter the world a second, third or even fourth time. It's a mother's nightmare.

It's also unsettling to think about the millions of little girls watching the commercial who will be counting the years until they can have their own little babies to play with. Unfortunately, the commercial doesn't mention stretch marks, postpartum depressions and spending the rest of one's life in control-top pantyhose.

As if we weren't having enough problems in our country with children, we now have misinformation, if not false advertising, about the sheer pleasures of childbirth.

For starters, real babies are much more complicated than a lump of plastic with a perpetual smile that wets it diaper once a week. Real babies are much more creative, and you can't store them in a toy box when you're bored with them or want to play a game of softball.

Mothers don't always look perky after giving birth. After each of our babies was born, it took at least three days of intensive therapy to remove the scowl from my face.

Ultrasound machines are vastly over-rated. In order to have a fuzzy, black-and-white photo of their unborn children, couples are now forking over more money than they spent on their wedding pictures or the deposit on their first home.

If doll-makers are seeking realistic role-models for little girls, I have a few suggestions. The first would be the "mommy's having a bad day and I'm staying out of her way doll". The doll's accessories would include a tiny, burned dinner, an alarm clock and a miniature pile of unpaid bills.

The "mommy's just stepped on the bathroom scale and is she ever upset doll" would also have a mommy doll's wardrobe in several sizes, including small, large and "don't leave the house".

The "mommy's been up all night with a sick kid doll" would definitely delay any premature desires to have a baby before one's time.

When it comes to pregnant dolls, reality can wait.

The Updated "Twelve Days of Christmas"

For the past 10 years, economists have followed the value of gifts in the popular Christmas carol, "The Twelve Days of Christmas." Evidently, the inflationary rate of the gifts is closely connected to our national inflation.

Why certain economists would choose to spend their holiday this way, rather than shop until they drop like the rest of us, is a complete mystery. However, they are pleased to announce that the total cost of the gifts is presently $15,760.70

That's right. That price tag includes everything from pricey "10 lords-a-leaping" ($3,012.63) to a bargain basement "partridge in a pear tree" ($34.99).

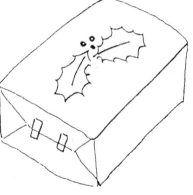

The good news is that the inflation growth from last year was only 1.1 percent. The bad news is that, even if you could afford to buy them, most of the items are politically incorrect for the 1990s.

Actually, that's not a big surprise. Most parents can't afford a nice family room until their families have grown up and left home. Our ability to buy fashionable clothing doesn't occur until we've lost our figures. And now - when it's possible to get a second mortgage on our home and buy the gifts in "The Twelve Days of Christmas" - it would be extremely unpopular to do so.

City fathers and writers of local zoning codes would undoubtedly frown upon a partridge in a pear tree, two turtle doves, four calling birds and six geese-a-laying in our backyard. If nothing else, we would fall out of grace with our next-door neighbors.

The purchase of three French hens would have all sorts of "buy American" people down our necks.

In light of the fact that trumpeter swans are a protected species, having seven swans-a-swimming would certainly result in a raid by the Department of Natural Resources's SWAT team.

If our living room had nine ladies dancing and 10 lords-a-leaping, we would probably be arrested for disturbing the peace.

Having eight maids-a-milking would most likely be more trouble than it's worth. Not only would the cows be a problem for the neighbors, but most of our leisure time would be spent filing reports for Social Security and state and federal taxes.

As for the 11 pipers piping - what kind of pipers are we talking about? Bagpipes, flutes or meerschaums? If it's the latter, we'll have to contend with the Clean Air Act.

Last but not least, there's the little problem with the 12 drummers drumming. Although we would face 12 counts of noise pollution and up to 18 months in jail, it would certainly cap off a Christmas we would never forget.

No Bones About It

Archaeologists are all a flutter about their latest find in northern Ohio. They have unearthed the bones of a third mastadon, allegedly killed by early North Americans at least 10,000 years ago. Their first clue was when they found flint flakes - sharp enough to serve as butchering tools - at the site.

While the discovery may indicate that some early Ohioans hunted without a license, it's difficult for many of us to share in their excitement. It's too late to arrest anyone. You must understand, of course, that everything I know about archaeology was gleaned from Indiana Jones movies.

In spite of the archaeology team's obvious enthusiasm, I believe that finding other artifacts, perhaps a tablecloth or a salt and pepper set, would have been much better proof of a prehistoric barbecue. I would associate human beings with a thousand other objects before I would think about flint flakes.

When I read about the discovery, I couldn't help having thoughts about my own mortality - thoughts that normally only occur before my first cup of morning coffee or when I walk into a dentist's office.

Imagine our worldly possessions buried for 10,000 years. Actually, that's not as long a time as you might think - some leftovers in my refrigerator have been stored and forgotten for roughly the same length of time.

Without knowing much else about us, what conclusions would future archaeologists make?

"An ancient religious site and an interesting fact about 20th century marriages have been uncovered by archaeologists in the area once known as the Upper Midwest.

"In the lower depths of a modest dwelling, they have

discovered an obscure shrine, apparently dedicated to the worship of fruits and vegetables. Clear jars of the vegetative matter were found carefully arranged on what appears to be crudely constructed shelves. Archaeologists are hoping to decipher one of the many labels attached to the containers: 'Brd & btt pckls 8/30/93'.

"The scientists have also determined that, contrary to popular belief, marriages of the late 20th century were not monogamous. In the recently unearthed dwelling, they discovered clothing in a storage area, belonging not to one woman but to several. Because the clothing belonging to obviously smaller women was stored toward the back of the space, they were also able to determine that the larger wife was the most dominant.

"In the deepest recesses of the storage space, the team also discovered an ancient device with numbers ranging from 0-200. The device appears to have thrown with reckless abandon into a corner of the cubicle by perhaps the larger woman.

"The most startling find was a one-wheeled vehicle, incapable of moving any distance. According to the latest testing methods, the spider webs attached to it are fully as old as the machine itself."

Face it. An old exercise bike would be much more dramatic than a handful of flint flakes.

Metric Mania

The metric system refuses to go away. According to the Federal Highway Administration, the dreaded metric conversion is inching forward and they hope to have metric signs on all of our highways by 1996.

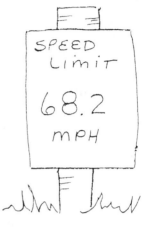

In other words, in two years few of us will know how fast we're driving or which traffic laws we're breaking.

In order to sweeten the deal for American drivers, the Association of State Transportation Officials has suggested rounding up the metric numbers to make the top highway speed 100 kilometers - or 68.2 miles - an hour.

I, for one, refuse to be swayed with promises of a higher speed limit. Of course, I might feel differently if I had smaller feet. It's almost a certainty that if our highways go metric, so will our sizes.

For more years than I care to remember, I have been plagued with large feet - size eleven to be exact. The doctor in attendance at my birth was heard to gasp when he saw them for the first time. By the time I was twelve, my feet were large enough to qualify for their own ZIP code. I'm sure there are remote islands in the South Pacific covering less area than my feet do.

It's a well-known fact that most shoe stores don't carry women's shoes larger than size ten. The only way to determine which do and which don't is to ask a salesclerk.

I usually hide behind a rack of shoes, hoping to avoid an embarrassing scene and hoping beyond hope that the salesperson passing by the rack will be discreet. With my face hidden in the turned-up collar of my trench coat I resemble a drug dealer more than an everyday shopper.

"Psssst! Excuse me, but do you happen to carry size eleven?"

"Do we . . .? Hey, Fred," he yells the full length of the store, "this lady wants size eleven!"

All heads in the store turn in my direction. Female customers, no larger than one of my thighs, pause as they try on shoes so small they would fit a Barbie doll. Women with feet so tiny that both of their feet would fit into one of my shoes gape in shocked disbelief.

Fred inevitably shouts back, "Eleven! Wow!"

Renting shoes in a bowling alley can also be a humbling experience. While other women can wear the cute, little rentals in pink and grey with pink laces, I'm usually stuck with the black, lunky men's shoes.

It could be worse. Under the metric system, my shoe size would be 42 and my other measurements would undoubtedly be in the thousands.

Although it might be exciting to drive at 110, I wonder what the Freds of this world would say about a woman with size 42 shoes.

Saving the Smallpox

As a country, we are divided into keepers and tossers.

So far, the greatest conflicts between the two groups have been whether to keep broken appliances on the basement shelves or whether the orange and green ashtray given to us by a distant relative should be sold at the next garage sale.

Perhaps the most serious showdown between the two groups occurred during the last week of 1993, when the World Health Organization (WHO), the tossers, expected labs in the U.S. and Russia to destroy the last smallpox virus samples in the world. In 1977 smallpox became the only disease ever eradicated.

Several groups of scientists, the keepers, have delayed that mutual destruction with a loud outcry. They reason that destroying the virus would mean the elimination of an entire species.

Sounding much like certain husbands when it comes to saving things, scientists think they might be able to use the virus again.

In the process, the scientists have become keepers and collectors, joining the ranks of people who save mounted butterfly or beer can collections and mothers who tape snips of baby hair and dried umbilical scabs in baby books.

There are similarities between the scientists and another group of collectors - mostly women - who keep smaller sizes in their closets, hoping to wear them again someday.

However, while most collectors are relatively harmless, keepers and vials of the smallpox virus sound like a deadly combination.

It's one thing for a person to show off a collection of

coins or stamps or even a baby book with discarded baby parts. It's quite another for a leading scientist to show off his prized collection at a social gathering.

"George, dear, why don't you show our guests your collection? Go ahead. Here are the frozen test tubes."

"Be careful, sweetheart. Those tubes are slippery. That nasty little virus has killed hundreds of millions since its first recorded attack in ancient Egypt. Now, hand them over carefully."

"Oops!"

That's the difference. Drop a butterfly collection and someone can get out the vacuum cleaner. Drop a few test tubes of smallpox virus and cleaning the floor is the last thing on people's minds. The very last thing.

The New Sensitivity

My on-the-street conversations with men are taking on a new twist.

Now that men are becoming more openly sensitive, our exchanges no longer consist of "How about those Twins?" or "Do you think we'll have more rain?"

Until recently, I was convinced that women had cornered the market on conversations about feelings. Men seemed less willing to bare their souls on busy street corners and in the aisles of grocery stores. They were forever checking their watches because they had places to go and things to do.

Ironically, now that men of the 90s are finally willing to share their innermost feelings, women are too busy to listen. Super Mom roles have taken their toll.

My simple "How are you?" in a grocery store this week, intended to go nowhere, led to a brief summary of a man's life. Obviously mistaking me for Oprah Winfrey or Sally Jesse Raphael, he described in painstaking detail his hobbies and interests, his hopes for world peace and the factors leading up to his high school G.P.A.

While I glanced at my watch, he began talking about his wife.

"You know, it might not work out. She's not the same girl I married."

I glanced at my watch once more. Barring any traffic problems, I should be able to drive home and have dinner on the table in 30 minutes.

I laughed nervously and the man sighed. As I headed for the checkout counter, it's almost a certainty that he muttered to himself, "Conversations with women are taking on a new twist."

His observation, "She's not the same girl I married", has haunted me all week. Of course, she's not the girl he married. Time has a way of changing all of us.

Imagine a middle-aged grandmother married to a man whose idea of a good life includes a drive-in movie, a cold six-pack and oversized, furry dice hanging from his rearview mirror. Or try to imagine a man returning home every night for 30 years to a sobbing bride and a meal of burned meatloaf.

Some changes are good. If not, my husband and I would still be sitting on Danish modern furniture, eating aluminum-wrapped TV dinners and playing canasta. I would still be unable to sleep with brush rollers in my hair and my husband would continue to sprinkle his sentences with "Cool, man" and "groovy".

Unfortunately, some good things were lost along the way - specifically, meaningful conversations in the frozen food section at the grocery store.

One Size Can't Fit All

One of the greatest myths facing the American consumer today is "one size fits all."

The manufacturers' premise seems to be that the same garment could fit any person, regardless of his or her size. In other words, I could share the clothes with Nancy Reagan.

If one size could truly fit everyone, it would also be possible to squeeze a 30-pound watermelon into a plastic sandwich bag. Placido Domingo and a racing jockey would wear the same size. It simply can't be done.

Like other fallacies, the "one size fits all" pitch can only lead to bitter disappointments and overdoses of chocolate.

Because of several clothing purchases made during the past few months, I'm convinced that I'm not the "all" the manufacturers had in mind.

First there were the golf socks. Labeled to fit sizes 7-11, a range of several inches, the heels retreated to my insteps within a few minutes after the first tee-off. It's almost a certainty I would have won the tournament if I wouldn't have been hobbling around like a geisha girl, whose feet had been bound at birth.

Pantyhose are also sold for multiple sizes. After relentless tugging at the waistband, out of the sight of witnesses, it's often my misfortune to discover the waistband has descended to the area of my knees. On more occasions than I care to remember, slipping pantyhose have caused me to waddle like Quasimoto, Notre Dame's literary bell ringer.

My latest purchase was a knit nightshirt with a label proudly proclaiming, "One size fits all". Feeling more like a full-body tourniquet, the shirt, worn in public, would invariably lead to the question no 51-year-old woman wants to hear: "When are you due?"

Sweat pants were the forerunners of "one size" fashions. After seeing them worn by people of all ages, I became convinced that the stretchy fabrics would cover my bulges. However, after purchasing several pairs and trying them on at home, I realized that even sweatpants came in limited sizes: small, smaller and "don't bend over".

The gray sweatpants made me look like a circus elephant on a bad hide day, and the blue ones gave me the appearance of a gigantic blueberry.

As a result of my confidence in "one size fits all" labels, I now have dresser drawers and closets full of never-worn clothes. When one of the "alls" shows up, she may have them.

Why Bowlers Smile

In his bestselling book, "The Government Racket", Martin Gross reveals surprising expenses in the 1992 federal budget.

According to Gross, the government spent $144,000 to study the relationship between pigeons and human economic laws and $500,000 to build a replica of Egypt's Great Pyramid in Indiana.

Those were minor expenses compared with the $2 billion spent for new furniture for Congress and government employees, and the $100 million needed each year to store helium gas underground in Texas. That's a lot of party balloons.

While all of these expenditures have been apparently justified, one government-subsidized study seems rather foolish. An undisclosed amount was doled out to study the cause of rudeness on tennis courts and the smiling patterns in bowling alleys.

The answers are quite obvious. The rudeness of tennis players and the bowlers' smiles are direct results of their proximity to food and refreshments.

Bowling alleys are carefully arranged so that no bowler has to be more than three feet away from food or drink. Each lane has a table with built-in cup holders, lights and ashtrays, and all of the lanes are close to the concession areas.

The table tops are perfect for holding bags of chips, peanuts and other basic food groups, as well as score sheets. Life doesn't get much better than spending time in a bowling alley.

Tennis courts, whether public or private, always seem to be located in remote, out-of-the-way places. Being conveniently located next to food and beverages is often sacrificed for landscaping and general appearance. For aesthetic reasons, you won't find beverage holders attached to the nets or insulated coolers incorporated into the courts.

Greasy hands created by potato chips may be great for creating a spin on a bowling ball, but the same grease wouldn't help a player's grip on a tennis racket. In fact, the very nature of tennis makes eating and drinking at the same time virtually impossible.

In between lobbing balls, chasing them down and wiping his or her forehead with wrist sweat bands, a typical tennis player is physically unable to carry around beverage cans, a box of Hostess cupcakes and chip dips. They simply don't have enough hands.

It's no surprise that bowlers smile and tennis players tend to be rude.

Controversy in the Bathroom

It's time to straighten out our priorities. While the rest of us have been worried about an unusual monsoon season and its impact on our rural economy, pollsters have been concerned with other problems.

We may be losing sleep over unplanted crops, uprisings in Europe and presidential haircuts, but the Northern Paper Company has been fretting about how people use bathroom tissue. Their survey makes us realize how petty and trivial the other problems really are.

In case you're interested, the survey of 1,200 people reveals that 40 percent crumple the paper when using it. Thirty percent wrap it around their hands and another 30 percent fold the paper carefully.

Not since a few years ago, when public debate was over whether bathroom tissue should roll toward the wall or toward the user, have people been so concerned with what goes on behind locked bathroom doors.

If the Northern Paper Company really wanted to serve the interests of their consumers, they should have asked a more relevant question. Why is it that no one wants to use the last square of paper on the roll?

I think it all has something to do with why people don't like to empty the last drops of milk out of a carton, but would rather place the container back in the refrigerator for the next person to find. It all has to do with replacers and non-replacers.

Replacers are given the dubious honor of filling empty gas tanks in cars. Non-replacers are guilty of leaving minute amounts of butter on butter dishes, rather than replacing what they've taken with a new stick.

But back to the bathroom tissue controversy. It's

none of my business whether people using our bathroom wrap the tissue around their hands or their heads, or whether their folding techniques resemble the Japanese art of origami. The only thing I can't figure out is why they can't reach into a cabinet a few inches from their knees and find a replacement roll to put on the dispenser.

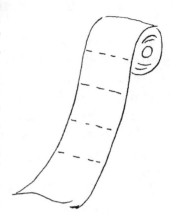

A tissue dispenser is hardly a complicated device with its tube and spring. In spite of its simplicity, it must appear threatening to people who can normally program VCRs and operate microwave ovens. Considering the fact that I'm the designated paper replacer, I've only had one bad experience with the simple mechanism.

One day, while I was paying more attention to my ranting and raving than to the task at hand, the metal tube flipped out of my hand and landed in the toilet bowl. It took all the courage I could muster to retrieve it.

Being a replacer isn't easy.

My Last Name is Andretti

I've become Walter Mitty. Although James Thurber's character became a military officer and a doctor when his mind went AWOL, I daydream about becoming a race car driver.

State laws may keep my road speeds at moderate levels, but they have no way of controlling the pit-stops, cheering crowds and high-speed curves in my mind.

I may look like any other middle-aged woman driving to the grocery store, but in my mind my last name is Andretti and the parking lot has become the Bonneville salt flats. While I may appear to be stopping for signal lights and following driving laws to the letter, spectators standing too close to the track are fleeing to safety and checkered flags are waving madly in the air.

My fantasies about race car driving came one gear-shift closer to reality last spring when I became the proud owner of a bright red sportscar. According to the manual, the car has a 24-valve, 222-horsepower, V-6 engine.

I have no idea what that means. The numbers may sound impressive to anyone who knows anything about cars, but as far as I know, V-6 sounds like a watered-down tomato juice. Because I don't really know how much power one horse has, the 222 doesn't mean much either.

Anyway, I've memorized the numbers, just in case anybody asks. I've never harbored fantasies about being a mechanic. It's just as easy to rattle off, "The suspension system has gas-charged MacPherson front struts" as it is to say, "How about those Twins?" when you know nothing about baseball.

Leaving the mechanic fantasies to other people, I

simply enjoy driving the car around, testing the buttons on the dashboard and listening to Paul Harvey over six radio speakers.

"The habit doesn't make the monk" is an old Spanish proverb. I might add, "A fast car doesn't make a race car driver."

While I always imagined that owning such a car would mean packing my bags and making new friends at Daytona, I'm still missing the vocabulary of a true racer.

One night this week, I had my first passenger in the car - my husband. It should be noted that his feet are firmly rooted in reality.

As we drove home from dinner, I tried to impress him as I banked curves and used the directional lights. As we pulled up behind a slow car, I decided to show him how I could pass, signal and maneuver the car like a professional.

He was clutching the dashboard as we passed the car and I gleefully shouted out, "Bite my tires!"

"I don't think that's the right expression, Carole. It's 'eat my dust', not 'bite my tires'."

Critics. As we drove home in silence, I wished that he could have seen me drive at Indianapolis and have listened to the shouts of the crowd. As my car crossed the finish line, no one questioned my vocabulary.

Better yet, he should heed the advice of Sam Levenson - "If your wife wants to drive, don't stand in her way."

You Are What You Eat

Courtship has become a complicated ritual.

While I was still in high school, shortly after the extinction of dinosaurs and just before the last glacier sighting, dating was simple. That included the pickup lines used by young men whenever the spirit or - in some cases - their hormones moved them.

"Duh, hey. Ain't you in my English class?"

The more polished males of the species would resort to approaches which were guaranteed to send chills up and down the spine of any blushing coed - "Hey, don't I know you?" or "You! Yeah, you! Your face looks familiar."

With lines like those, it's no wonder we all rushed into early marriage.

Since those days in the 1950s, courtships became big business. Prospective daters began taking out classified ads: "Handsome SWM seeks someone special who can mountain-climb, skydive and cook three meals a day."

They began exchanging video tapes through dating agencies.

"Is the camera running? Okay. Hi, my name is Babs. I'm a Libra. I like taking long walks on the beach, watching fires in the fireplace on cold, winter nights and.....um.....could we start this over again?"

Zodiac signs became calling cards.

"Hi, what's your sign? You're a Leo? How wonderful! I'm a Capricorn, and right now my moon is in its last cusp, my sign in the seventh house and I have no idea what I'm talking about. Hey, ain't you in my English class?"

It now appears that Zodiac signs have run the way of hula hoops and Frankie Avalon 45s. According to the latest issue of Omni magazine, the trendiest pickup line isn't "What's your sign?" but rather, "What's your fruit or vegetable?"

Apparently, the old axiom is true - you are what you eat. The Smell and Taste Treatment Foundation in Chicago argues that passive personalities will choose applesauce over fresh apples and creamed corn over corn on the cob. However, the opposite choices would be made by "an aggressive go-getter who works hard, plays hard and won't take no for an answer."

The theory also tries to explain why aggressive people prefer oranges, bananas and apples, while sensitive souls choose eggplant, corn and tomatoes.

If this new method of finding the perfect person continues, anyone hoping to have a successful social life will have to study nutrition or read cookbooks. By mentioning a socially taboo food, such as mashed potatoes or strained carrots, an otherwise nice person could be doomed to a life of misery.

My favorite vegetable? Potato chips, of course.

Good News for Pear Shapes

It's not a simple case of comparing apples with pears.

When researchers at the University of Minnesota refer to fruit these days, they're talking about women shaped like pears and women shaped like apples.

According to a study, pear-shaped women, ages 55-69, with hips considerably larger than their waists, live the longest. To be more specific, a woman's waistline should be 76 percent as large as her hips.

While the researchers may refer to the project as a scientific study, I think of it as a personal affront. In one fell swoop, they've managed to focus on my least desirable measurements. The only worse thing would have been to have my name mentioned in the study.

I wonder why the ratio had to deal with the waist and hips. Why not look at the index finger and the wrist, or the ankle as opposed to the circumference of the neck?

It's not fair.

I strongly suspect that, in order to decide which body parts to compare, researchers took turns spinning a large roulette wheel, labeled with body parts rather than numbers. Taking a cue from "Wheel of Fortune's" Pat Sajak, one researcher gave the wheel a spin as others looked on.

"Around and around it goes, ladies and gentlemen. It's slowing down! It's the forearm! No, it's the ankle! Wait a minute! It's the waist!"

On the second spin, the wheel stopped on "hips".

"That's it," exclaimed the researcher-turned-gameshow emcee. "We'll compare waists with hips! Only one decision remains: How much smaller than the hips should the waist be?"

"Sir?" All heads turned to the back of the lab, where a grants-writer stood meekly by himself. "How about 76? It has a nice patriotic ring to it. It's bound to be a shoo-in for a hefty government grant."

The rest is history. Little thought was given to the fact that cargoes naturally shift in woman 55 and older, or that those women consider elastic-waist slacks to be the eighth wonder of the world.

After reading about the study, I decided to take my own measurements. Outside of a six-inch ruler, the only measuring device in our house was a steel measuring tape, normally used to measure floors in rooms and other flat surfaces.

Unfortunately, I'm round rather than flat.

As I stood in the kitchen and tried to wrap the inflexible metal strip tightly around my hips, the tape ensnared two kitchen chairs and the microwave oven. After wrestling with the measuring tape for what seemed like hours, I finally determined that my hips measured six feet and two inches.

I no longer worry about longevity. When my husband finds his measuring tape, now twisted and bent beyond recognition, my days will be undoubtedly be numbered, anyway.

Love Thy Neighbor

One of the greatest deterrents to crime while I was growing up was my mother asking, "What will the neighbors think?"

Whether it was a bike left on the front sidewalk or returning home too late after a date, my mother was quick to throw the neighbors in my face. What "they" would think on any given occasion pretty much determined the extent of my activities and social life. Their opinions mattered.

Even though dates and bicycles are only dim memories in the cobwebbed recesses of my mind, I still worry about the neighbors. By picking up a newspaper or tuning into a news program, you can find them on a regular basis.

Whenever a newsworthy crime is committed, journalists feel compelled to interview the suspect's next-door neighbors.

It should be noted that what may be a crime in one area might not be in another. To illustrate my point, let's look at a crime in my hometown and how the neighbors might react.

"Good evening. I'm interviewing the next-door neighbors of Lars Larsen, who recently defied local government by placing three empty tuna cans in his regular trash. As you may know, cans of any type are recyclable and should be placed in separate containers. We're talking to John Johnson, who has known the defendant for 20 years. What was your impression of Mr. Larsen, Sir?"

"Oh, he was the sort of guy who always kept to himself. He seemed okay, I guess."

Whether the offense was returning library books past their due date or hijacking a bus full of school children, the response is always the same. Or, is it?

Given the fact that some of us don't always keep to ourselves, what will our neighbors say when faced with a video camera and bright lights?

"Well, I guess you could call her forgetful. Sometime we would see her get in and out of her car three or four times before she would actually take off. As she would run back and forth from the car to the house, we would hear her muttering about her checkbook, the grocery list and the car keys on a regular basis."

"You know, I always wondered what she fed her family. The garbage bags were always filled with carry-out pizza boxes and little else."

"She didn't spend much time outside. I guess the last time we talked was when she last cleaned the windows, maybe 1986 or 1987."

"Which one are you talking about? The skinny woman or the woman who's not so skinny? They look a lot alike and my wife insists they're the same person. Personally, I don't believe one woman could gain and lose that much weight during a few years."

The lesson to be learned by all of this is quite simple - love thy neighbor. You never know when they might be asked for a character reference.

Safe Bombs - Making our World a Better Place

We may all sleep more soundly at night. Our country is building safer bombs.

I know - safe bombs may sound like an oxymoron - but let me explain.

For several years now, the civilian population has been spending its leisure hours recycling newspapers, glass, plastics and aluminum and doing its level best to improve the environment. We eat politically correct tuna. We don't burn fires. We worry about the ozone layer although we've never seen it. When a volcano spews ash into the atmosphere, we take it personally.

Not to be outdone, the federal government, prodded by the Environmental Protection Agency, has decided to build better bombs.

The Wall Street Journal reported last spring that the U.S. Air Force plans to outfit its nuclear missiles with cooling systems that don't use chlorofluorocarbons (CFCs).

For those of us who don't know the difference between a CFC and a VCR, CFCs are blamed for depleting the atmosphere's ozone layer, leading to skin cancers, glaucoma and a long list of other diseases.

In other words, each intercontinental ballistic missile will still be able to wipe out entire cities, but the people won't have to worry about sunburns or their eyesight. The situation reminds me of one of my grandmother's favorite sayings, "It's like throwing the baby out with the bath water."

If the government is truly serious about the future of our planet, perhaps they should have given some thought to eliminating the bombs completely.

The very least they could have done was eliminate the triplicate and quadruplicate paper forms they keep generating. By reducing tax instruction booklets to less than best-seller novel size, they could save several oxygen-producing forests.

I suspect those ideas are too simplistic. While the rest of us scrape labels from plastic bottles and lug sorted recyclables out to the curbs, the government clearly marches to a different drummer.

Compromises might work. Perhaps if we would all agree to stop drinking out of styrofoam cups and using antiperspirants from aerosol cans, they could work out this bomb-fetish business. Contests could be held nationwide to determine uses for empty missile silos. They could possibly be used as pre-dug landfill areas.

If the government thinks outfitting the nuclear missiles with different cooling systems is the answer, our only option is to continue recycling - as if there were no tomorrow.

The Fitting Room Game

My figure happens to be very flexible. In fact, there are no sizes I can't squeeze into.

Salesclerks in clothing stores have learned to take my clothes fittings in stride. I've seen them wince when I've tried to wedge a size-11 foot into a size-eight shoe. I've watched them grimace as I've stretched 28-inch waistbands over hips twice that size.

Most salesclerks subscribe to a common theory - the customer is always right. That's why they don't blink an eye when I ask for something little in a size ten. Not too loudly, mind you, but loud enough for all customers to hear me from housewares to children's clothing.

They seem to instinctively know that I will eventually come to my senses and realize that the pair of slacks will only cover one of my thighs.

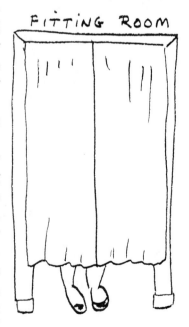

They're very patient as I play the clothing game.

"Excuse me," I'll whisper from behind the dressing room door. "These tens are the teeniest bit tight. Maybe I should try a larger size."

At this point in the game, a salesclerk with any business sense at all will respond, "Oh, dear! You know, that brand tends to run quite small on everybody." Without another word, she will slither discreetly through the department and bring back something "slightly" larger. Perhaps a size 14.

She automatically loses game points if she refers to the amended size. If she plays the game right, she earns the usual commission.

It's a game where no one loses.

Last week I found a salesclerk who wasn't attuned to the finer points of the game. Unfortunately, it happened during a trip to the newest clothing store in our hometown and the salesclerk knew me by name.

I had just selected a birthday gift for our youngest daughter when she turned to me and asked, "Will there be anything else?"

She must not have realized that those were the opening words for the clothing game.

"I think so. Do you happen to have long, full summer skirts?"

Point. Rebound. We were on a roll. In seconds I would be in the fitting room and the play would be complete.

Instead of showing me the various styles in tiny sizes - the ones you find first on clothing racks - she chose to forfeit the game.

"Oh, I'm sorry. We don't carry your size here. We only have junior and misses sizes. We don't carry women's sizes at all."

The final score flashed before my eyes: Customer - one, salesclerk - zero. Somewhere in the recesses of my mind a crowd booed.

After paying for the original purchase, I should have gone home, but we were suddenly playing overtime and the final point was at hand.

"But I don't wear women's sizes," I lamented. It was still possible that she could claim temporary insanity, or at least temporary stupidity. In that case, the play would continue.

"Oh." Her eyebrows arched. "You don't?"

At least I learned an important lesson that day: If you don't know the rules, don't play the game.

Hostile House Plants

Remember laughing at the movie title, "Attack of the Killer Tomatoes"? Were you guffawing with friends when the man-eating plant in "Little Shop of Horrors" made its first appearance?

We may have laughed prematurely. Two articles indicate that plants intend to bite the hands that feed and water them. In the first article, the Consumer Products Safety Commission reported that eight people a day in our country blame their injuries on house plants.

If the report is to be taken seriously, it's very possible that a tiptoe through the tulips and a walk down the primrose path aren't what they're cut out to be. Stopping to smell the roses may prove dangerous.

In recent weeks, the U.S. House approved spending $1.1 million for a plant stress lab at Texas Tech University. Instead of nipping the threats from plants in the bud, we are now faced with the further insult of paying for their stresses and neuroses.

Throughout all of this controversy, we must keep in mind the eight citizens who are blaming their daily injuries on house plants. Do the plants furtively edge over to the tops of stairs, causing their owners to fall? Are there unknown hostilities in African Violets?

Let's face it - we are the hapless victims of flower power. Plants have some of the strongest voices in Washington. Waving the emotional banners of rain forests, redwoods and certain "endangered species", plants have enlisted the aid of Vice President Gore and several well-known Hollywood actors.

While we are being distracted with those noble causes, plants continue to maim citizens on a regular basis. We shouldn't dismiss the flora felons because of their diminutive sizes. Consider the world's largest flower, the Rafflesia, which reaches a diameter of 36 inches, weighs up to 15 pounds and grows on the mountain slopes of Borneo.

If the Rapplesias are ever exported, florists should seriously consider changing their motto to "Slay it with flowers."

Even without the above information, I have acquired a new respect for the stamina of house plants. Last week, I returned home after a stay at the lake, fully expecting my neglected house plants to look like dried flower arrangements. Imagine my surprise when I found out they had literally taken over the house.

Their tendrils were gripped around faucets in the kitchen and bathroom. Without a machete I found certain routes through the house completely impassable. Except for the obvious absence of parrots and snakes, our living room resembled a South American jungle.

In light of the daily accidents caused by plants and their amazing ability to thrive without water, the $1.1 million might have been better spent elsewhere.

Building a Better Man

Because she built a better man, a California book publisher is hoping people will beat a path to her door. Barbara LesStrang is selling handsome, male mannequins - called Safe-T-Men - to sit in cars and scare away would-be carjackers.

Much to her surprise, her lifelike beefcakes have been selling like hot cakes.

Although rural residents of the Upper Midwest need protection from carjackers as much as they need hurricane insurance, lifelike mannequins might prove useful in other ways. In fact, I'm seriously considering ordering one adult male version and several child-sized models.

The current best-seller mannequin, a male, is the strong, silent type but he has definite drawbacks. Not unlike the models already available in our area, he is unable to move a muscle or blink an eye. Propped in a recliner chair with a TV remote control, he would be indistinguishable from what many of us already have in our homes.

The male model I have in mind would be more of a 90s guy. I would prop him up against the kitchen sink, in full view of the neighbors, who would gasp, "Look! He's doing the dishes!" At other times, he could be balanced next to the stove, ostensibly preparing seafood fettuccine or spinach quiche.

I would be the envy of the neighborhood.

Although he would be unable to share his innermost feelings, he might be programmed to recite a patter of pleasantries - "Gosh, you look great. Have you lost some weight?" or "Next to you, Julia Roberts would pale by comparison."

The child-sized models would serve another purpose. Placed in a neat row in the back seat of my car, they would be used as surrogate grandchildren.

On my frequent trips to the local grocery store, invariably someone, between the produce department and the canned vegetables, will inquire, "Do you have any grandchildren yet?"

Instead of mumbling feeble excuses about career-oriented, fast-track daughters, I would exclaim, "Do I? It just so happens I have all eight of them in the back seat of the car! They're all so bright! Would you like to meet them?"

I'll just have to hope that the interrogator has melting ice cream in her grocery cart and will offer her excuses. Actual face-to-face contact with my plastic-faced grandchildren might prove to be an unsettling experience.

Either way, the questions would stop.

I'm Younger Than I Look

When a person becomes a senior citizen - if you'll pardon the expression - is a gray area.

Depending upon the restaurant franchise or the particular store, discounts are heaped upon persons who have turned 55, 60 or 65. Determining whether a person is a senior citizen is apparently a capricious matter.

Last week, when I least expected it, I crossed the thin line between middle age and senior citizen. I became my mother.

It all began with a simple purchase at one of the local drugstores. I waited patiently in line to buy a card for my sister ("It'd be really nice hearing from you . . . Unless you're dead. Then it'd be kind of scary").

I mention the card's message as evidence of the mood I was in that brisk, sunny, winter afternoon. Wild. Carefree. Happy I had made it through the day without a run in my pantyhose. Life didn't get much better.

After all, it wasn't as though I was standing there hoping to buy a year's supply of fiber laxatives or geriatric vitamins.

"Will that be all?" asked the young blonde at the cash register. As I nodded, she cheerfully added, "Will you be using your senior citizen discount?"

As her words sunk in, my world collapsed. Instead of entering a drug store, I had mistakenly entered Rod Serling's Twilight Zone. I tried to think of a snappy comeback to set the salesclerk straight, but regardless of how many times I opened and closed my mouth, no sounds came out.

"I...I...I'm only 50", I finally blubbered. Although I shouldn't have been surprised, a hot, salty tear rolled down my right cheek. Melancholy songs on the radio often evoke the same response.

"I'm so-o-o sorry," she apologized as she rang up the sale.

Now that a few days have passed, I'm able to view the situation more objectively. I'm also convinced that people who ask "Will you be using your senior citizen discount?", before one's time has come, should be exiled to a remote island.

Left to their own resources, they could cavort on the beach with people who ask "When is the baby due?" when you're not pregnant, and people who can't end conversations without shouting, "Have a nice day!"

There are times when "I'm sorry" simply doesn't cut it.

One friend, upon hearing my story, asked, "Well, did you take it?"

"Take what?"

"The discount! Did you take it?"

All I could do was smile - and imagine him on the remote island with all the others.

Recalling the Dishwasher

I'm faced with a most difficult decision. Will I wash dishes by hand or will I let my kitchen go up in roaring flames?

A short time ago, a large catalog retailer - which will remain nameless - decided to recall 300,000 dishwashers, sold between February 1990 and October 1992. If my memory serves me correctly, I was introduced to our dishwasher during that time.

The recall evokes mixed emotions. Should I face the ravages of dishwater hands and a cluttered kitchen counter just so a few company executives can rest easier? What are the chances that the company - which recently issued its last catalog and closed several stores - will get its act together and replace my dishwasher before my 90th birthday?

According to the recall notices, overheated timers in the dishwashers may lead to disastrous kitchen fires. Actually, I run that same risk everytime I prepare a meal. I've set off the smoke alarm so many times that meals without the buzzing in the background seem to be missing something.

It's not that I have an appliance fetish. Under similar circumstances, I would gladly give up the dust-covered hot rollers on the back shelf of the closet or any of the three electric bun warmers received as wedding gifts. I would gladly surrender my Veg-a-Matic.

My dishwasher ranks right up there with chocolates and lingering bubble baths. I'll never forget the day we met.

She arrived during a low point in my life. Her predecessor had met an "untimerly" demise during a second rinse cycle. Within seconds, I placed the equivalent of a 911 call to the local catalog store.

"Help!" I shouted into the phone. "My dishwasher died and I need a replacement!"

"This sounds rather urgent," noted the dealer.

"Urgent? I might go crazy without a dishwasher!"

Apparently, words like "crazy" aren't taken lightly in catalog stores. Knowing a damsel in distress when he heard one, the dealer had the situation under control and promised to call the Minneapolis warehouse immediately. On the following afternoon, the dishwasher would be delivered to my doorstep.

Mere words can't fully describe the subsequent 28 hours. With the knowledge that a new dishwasher was on the way, dirty dishes were piled on every available counter space for the next four meals. By the time the phone rang on the following afternoon, the mountain of egg-crusted forks and spaghetti-splattered dishes had reached such magnifi-

cent proportions that I doubted seriously whether I would be able to find the phone.

"We're on our way!" It was the voice of the dealer, my knight in shining armor.

"This must have been quite an emergency", noted one of the delivery men. He nodded at the packing box, which was plastered with neon-bright labels - "Urgent!" "Rush!" "One Day Delivery!"

To add insult to injury, the same company which recalled our dishwasher has now decided to close our local catalog store.

Appliance emergencies will never be the same.

Chilling News

Here's the perfect contest for people who are tired of hearing the question, "Is it cold enough for you?"

OMNI magazine is sponsoring a contest in which the winner will be placed in cryonic suspension - frozen - after death, with the possibility of being revived in the future. According to the contest rules, the dubious winner has to write an essay, 250 words or less, telling why he or she should be selected for possible revival in the future.

The catch word here is "possible". It's also possible that the winner will become the first victim of freezer burn, a questionable claim to fame.

There's no mention of what happens to runners-up in the contest, but they will most likely be forced to spend their remaining years in Thief River Falls or Fargo.

On the other hand, winning the contest might offer several benefits.

For one thing, after being frozen at subzero temperatures for 200 years, I might learn to appreciate how mild Midwestern winters really are. A windchill factor of 20 degrees below zero would seem like child's play.

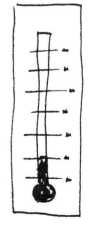

It would be nice to outlive the leftovers in my refrigerator and the plastic garbage bags in the county landfill.

I've never wanted to be a burden to my children in my later years, and cryonics might be the answer. Instead, I could be burden to my great-great-great-grandchildren, who would be complete strangers.

It might be fun to come back and watch the Twins win their 50th World Series.

For the first time in my life, I would look younger than my actual age.

"You're 250? You don't look a day over 80!" Of course, at the rate it's going, the English language will be so corrupted that the question will sound more like, "Yure toofifee? you donluke a da oover ayee!"

As the oldest woman in the world, I would have certain privileges. No one would dare tell me to floss my teeth or how much I should weigh. Everything I would say would seem profound.

"Hark!" the android would shout, as he calms down the curious crowd. "The old one speaks." To throw them off, I would throw out archaic phases: "Rock and roll is here to stay" and "What is your ZIP code?"

At the current inflation rate, it might be interesting to wake up after a couple of hundred years and buy a loaf of bread for $20,000.

If it's true that all things come full circle, it's also possible that I might be revived when full-bodied, corpulent women are the rage. As the head scientist, Dr. Thor, supervises my final thaw, he would be heard to exclaim, "(Gasp!) She's fat. (Gasp!) She's beautiful!"

I think I'll enter.

Getting Down to the Root of the Problem

The man wore gloves and a mask.

For one split second I wasn't sure whether he was going to take all my money or do root canal work. As it turned out, he did both.

While most people were happily counting off the days until December 25, I was approaching December 22 - my date with dental destiny - with dread. Having root canal work wasn't my idea of holiday entertainment.

When my regular dentist referred me to the out-of-town specialist, he told me it was for my own good. I strongly suspected that he and the local chamber of commerce had determined that my usual bloodcurdling screams would discourage holiday shoppers in my hometown.

It's a well known fact that I have a very low pain threshold when it comes to dentists or giving birth. It's so low that I need a general anesthetic before even calling a dentist for an appointment.

After I had been guided through a maze of hallways to the examining room, the masked dentist carefully explained that my broken tooth had left me with three options in life. I could end up looking like one of the toothless crones in the opening scene of

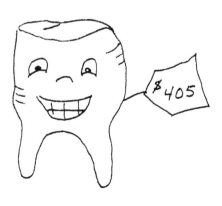

"Macbeth", I could spend the equivalent of the national debt on a partial plate, or I could have the root canal work.

Faced with a no-win situation, I chose the latter.

"The procedure will cost about $405," he added. The "about" was an unusual touch. It was like saying Pasadena, CA has "about" 118,072 residents.

As I mentally pondered the mysteries of uneven numbers, the dentist went about his business.

I don't wish to ruin the plot for anyone who hasn't experienced root canal work. Suffice it to say that enough wire was run through my tooth to link phone lines between two continents. On a pleasure scale it ranked right down there with being flattened by a steam roller or crawling on hands and knees through broken glass.

When it was all over, I was given a temporary filling.

"Don't chew anything on the left side of your mouth until February and your next appointment," admonished the masked man.

"But this is Christmas!" I whined. "What about the foods I've prepared?"

"You don't need them." He snapped off a latex glove for dramatic emphasis.

Now, telling me not to eat is like telling a woman in advanced labor not to push or advising a pyromaniac not to play with matches.

I tried my best to follow his instructions. With the holiday season well underway, I turned down offers of peanut brittle, nutty fudge and other seasonal treats. I stoically refused to bite into decorated Christmas cookies.

After what seemed like an eternity, I finally yielded to temptation. Life is too short, I reasoned.

The caramel was delicious. The filling lasted three hours.

Happy About Desserts

Health magazine recently bestowed the "What'd Ya Do With the Grant Money Award" to Stanford University.

Not knowing whether to spend the money studying global hunger or world peace, Stanford decided to study why certain people enjoy eating desserts. Without a PhD of my own, it's hard to follow their logic.

The study found that when happy people choose desserts, they are usually happy with their choices, and that when unhappy people make their selections, they are usually unhappy. Surprise, surprise.

If there's anything more pathetic than grant money spent on studying desserts, it's eating with a person who gets no thrills from devouring them.

In my opinion, desserts are as important as any other basic food group. Protected by the Bill of Rights - "life, liberty and the pursuit of chocolate" - desserts are much more pleasurable than plates of collard greens or limp spinach.

Going out for dinner with someone who doesn't share my enthusiasm for desserts isn't a pleasant experience. Diving with reckless abandon into a serving of pie a la mode can be a guilt trip when the other person caps off the meal with a glass of water.

"Oh, I can't eat another bite," she laments, as she finishes her three leaves of lettuce and clump of cottage cheese. "But why don't you go ahead and order something?"

Right. Like I'm going to order the Triple Chocolate Brownie Surprise while she sits there, looking like a poster child for anorexia.

Another moment of anguish arises when the dessert cart rolls around and she suggests, "Why don't we split a dessert?"

Sharing a dessert ranks right up there with sharing a husband. It's not a happy scenario.

When someone offers to split a dessert, I'm left with three choices. For starters, I can simply blurt out, "I'm sorry, but I'm going to inhale the whole thing, including the plate."

The second choice is to shrug my shoulders and play the role of a martyr as she hands over a paper-thin slice of Black Forest Torte.

When the first two choices fall by the wayside, I simply become belligerent and tell the waiter, "I can't make up my mind. Give me one of everything on the cart." For added emphasis I jump on the table and shout, "Life is too short!"

Some people have no sense of fun when it comes to eating dessert. To them it's just another tedious activity, like changing the paper in a bird cage. They will never know the thrill of having a meaningful relationship with a Turtle Sundae. How I pity them.

Pampered in Somalia

Until now, manufacturers of disposable diapers have met the needs of baby girls and baby boys. If our military maneuvers in Somalia continue, they may soon have two new sizes, "Enlisted Men" and "Commissioned Officers".

According to an NBC cameraman who returned from the Front, U.S. troops will reportedly trade almost anything for a nice box of disposable diapers. With water in short supply, soldiers have found the products perfect for cleaning up after a hard day's work. When slightly moistened, the diapers provide a great rundown, a close second to a real bath.

The cameraman noted, "They leave you smelling baby-powder fresh."

Although the practice presents an incongruous picture - a rough and ready soldier smelling like a baby's bottom - it suggests that we might better send our troops paper diapers rather than chocolate chip cookies.

In the wildest stretch of my imagination I can't envision a John Wayne war movie with the hero asking, "Does anyone have a Huggie?"

If this disposable diaper trend continues, it's also possible that CO's in wartime will no longer be conscientious objectors, but rather cloth objectors. In this case, it's almost a certainty that anyone protesting the use of disposable diapers on battle fields would also be opposed to defoliating trees and creating open fires. Given past experiences, I doubt whether the cloth diaper vs. the paper diaper controversy is given a high priority in the Pentagon.

Marine recruiting posters will have to be reworded: "All we're looking for is a few, good, sweet-smelling men."

For that matter, military recruiting methods will also be upscaled.

"To sum it all up, son, there are definite advantages to joining today's Army - on-the-job training, a paid college education, predictable meals, a shiny gun and free uniforms. Any questions?"

"Yes, sir," stammers the young recruit. "Those things are all very nice, but will we be issued regular or nighttime Pampers? I really get dirty when I'm outdoors a lot."

Gen. William Westmoreland once stated, "They're asking women to do impossible things. I don't believe women can carry a pack, live in a foxhole, or go a week without a bath."

Given an ample supply of disposable diapers, anything would be possible, General.

For the first time in our country's history, our fighting forces will be associated with a word rarely heard in military ranks - pampered.

Guess Who's Not Coming to Dinner?

Unless your name appears on an exclusive database of 1,000 names, not even Karl Malden and a fistful of credit cards could get you into Prive (pronounced pree-vay), located on Manhattan's Upper East Side.

One of the restaurant owners, Michael van Cleef Ault, noted, "We want people who are chic, high-energy, smart, sexy - we want a scene."

In an article in the New York Times, he added that he is seeking "people who are fun and some people are more fun than others."

All right, we can take a hint. Evidently, the Prive isn't ready for our own Midwestern chic - those of us who keep our pickups in working order and know most of the lyrics on the country-western top 40 charts.

Unlike the Prive, which will only cater to the owners' personal friends from the Newport-Palm Beach-Paris circuit, our small-town cafes will serve anyone. If it's high-rolling fun they're seeking, customers can always shake dice in a cup and risk buying coffee for everyone at their table.

Another nice feature about small-town cafes is that they don't stand on formality and strict dress codes. Although a man's favorite brand of truck may be an extremely personal matter, not unlike his choice of religion or after-shave, it's not uncommon to see a

Ford cap across the table from a Chevy cap. People in our part of the country may have their differences, but that doesn't keep them from sharing pie and coffee at the same table.

Our cafes will even serve men dressed in suits and ties - although their getups are dead giveaways that they're in town attending a funeral or a wedding or they're out-of-town bank examiners.

As far as world travels go, cafe patrons don't really care if you've been to Newport or Palm Beach or if your idea of a fun time is a car trip to the state fair. In fact, there's an unwritten rule about not bragging about trips you've taken outside the county. If you're a snowbird, returning home with a deep tan in the middle of March, you should be prepared for some good-natured kidding.

When it comes to fun, a hometown cafe is the place to be. If one of the customers is seen driving around town in a new car, or is seen on his yard with a new lawn mower, the fun never ends.

Perhaps the only thing that could go wrong in a cafe would be if one of Mr. Ault's "chic, high-energy, smart and sexy" customers would show up. While the rest of us would gawk, someone would invariably dial 911.

While the owners of Prive (pronounced pree-vay) may think their restaurant's exclusive list is a smart idea, I tend to think it's an idea destined for the privy (pronounced pri-vee).

Chocolate Testing

The President's latest cuts haven't been everyone's cup of tea. I refer, of course, to the U.S. Board of Tea Experts, one of 700 advisory boards and commissions scheduled for the budget ax.

Since the Tea Importation Act of 1897 the board has had a nice thing going. Sequestered in the New York offices of the Food and Drug Administration, the head tea taster, spittoon in lap, has been sampling 300 cups of tea a day and receiving $68,000 in annual pay.

For $68,000 a year, I would gladly sip, swish and spit 300 times a day. I'm already sipping and swishing 300 cups of coffee a day with no compensation.

If the President is really serious about changes in this new administration, I would propose that the tea board be replaced with a Board of Chocolate Experts. Let the few consumers of foreign teas run their own risks.

With apologies to David Letterman, I would like to offer the top five reasons why this board should be established and why I should be named the country's top chocolate taster:

1. Chocolate has been one of my best friends for most of my life, so costly on-the-job training would be unnecessary. I don't have to be convinced that chocolate is one of the basic food groups. If it's true that all matter can be classified as animal, vegetable or mineral, chocolate is definitely my favorite vegetable.

2. Office expenses could be cut with the elimination of the unsightly spittoon. Biting, chewing and swallowing chocolate would be much more civilized than the present system.

3. On second thought, why not eliminate the costly

offices altogether? I would gladly taste test 300 samples of chocolate a day in my own kitchen.

4. The creation of the board would lead to few cries of "pork barreling". As far as I know, "chocolate barreling" is politically correct.

5. Finally, the best reason of all - I wouldn't have to be paid for this service to my country. I would even pay postage for the samples delivered to my door.

It's time for our country to pay less attention to exotic Keemun, Assam, Oolong and Darjeeling teas and pay more attention to what really matters - the nougats, the caramels and the cream centers.

While the talk everywhere these days is about belt tightening, I would be willing to unbuckle mine in the interest of safe chocolate.

Along with chocoholics everywhere, I offer these final words to departing members of the Board of Tea Experts: S-Oolong, it's been nice tea know you.

Body Piercing

I'm afraid of having bad weather this winter.

What worries me most isn't having my eyelids frozen shut or having my tongue stuck to a metal car door. It's having my body pierced in places normally only mentioned in medical journals or Gray's Anatomy.

The latest fad in Los Angeles and Manhattan is body piercing. No longer content to have metal hoops dangling from their ear lobes and noses, trend setters are now having other body parts pierced, including navels, eyebrows and tongues.

And what, you may ask, does this body piercing business have to do with blizzards? There's a logical connection. Of course I've also been able to find logic in grocery store shelf arrangements and pantyhose sizing.

About 10 years ago, I was stranded in Mankato during a blizzard. It might have been winter, but in Minnesota a person can never be sure.

Unable to travel home from Minneapolis, I accepted the situation with unusual grace and charm. After checking into the only available motel room, and unable to find solace in any of the 31 cable TV channels, I went malling.

The plastic cards in my purse were still cool when I spotted the sign in the mall's hallway: "Ears pierced - While you wait." Partly out of curiosity - and partly because I wondered if ears could be left for piercing and then picked up later - I wandered into the nearby jewelry store mentioned on the sign.

Whether it was because of the storm raging outside or the barometric pressure playing tricks with my mind, it was an uncharacteristic move. For two decades I had been the person who proclaimed, "I'll never have my ears pierced."

Never happened in Mankato, on a day marked with waist-high snow drifts and bone-chilling temperatures. By the time I left the mall, my ear lobes were red, swollen and very much pierced. It might have been my imagination, but I was almost certain that the metal earrings were picking up faint signals from a radio station.

After reading about body piercing this past week, I found myself saying, "I would never have my navel pierced." Of course, the weather has been unusually mild and I haven't been in a mall since last spring's big thaw.

With my luck, there's a sign out there, reading "Navels, eyebrows and tongues pierced - While you wait."

The only thing keeping me from resembling a human pin cushion is unseasonably warm weather.

The Garden that Refused to Die

After frost hit our garden last fall, my first words were "Free at last!"

For three months, my life resembled a remake of the movie, "Attack of the Killer Tomatoes". My version could appropriately be called "The Garden that Refused to Die."

After I had spent two months hoeing, pulling weeds, fertilizing, staking plants and praying for sunshine, my garden finally yielded its first ripe tomato in mid-July. With the excitement normally reserved for first dates or lottery wins, I proudly showed off the first fruit of my labors to everyone in sight. Even the trashman and the mailman were suitably impressed.

A few weeks later, in spite of the unusually cold weather and monsoon rains, I discovered a second tomato. Although math has never been one of my strong points, it became quite clear that, at the present rate, I would be lucky to keep my family fed beyond a couple of meals. By squirrel standards, I wasn't doing very well with storing food for the winter months.

As my garden took on a personality of its own, it also became increasingly stubborn. It had its own agenda.

It may have been my imagination, but on quiet summer nights I was sure I could hear muffled shouts from the garden: "Hey, no, we won't grow! Hey, no, we won't grown!"

By the time I had given up, the garden became a wild profusion of yellows and reds. I suddenly had enough tomatoes to feed the western hemisphere.

My original goal had been to can two or three dozen quarts of tomatoes. Four hours into the first tomato attack, that goal had been met and more tomatoes than I had ever seen before were stockpiled outside the kitchen door.

Succumbing to the feverish forces of tomatomania, I tried new recipes and invented others. Bored with stewed tomatoes, I tried salsa, spaghetti sauce, tomato juice and chili sauce. For several weekends, from sunrise to sunset, preserving tomatoes for the long winter ahead became my obsession.

Every canning jar in the house was commissioned into active duty as the garden spewed forth more and still more tomatoes. Ironically, the more I canned and gave up other responsibilities, the closer my family came to the brink of starvation.

By the time the last shelf in the basement had been filled with jars, I began dreaming about an early frost.

If people can make plants grow by talking to them, I figured that threats might make them stop producing.

"Stop! Enough!" I shouted at the prolific plants.

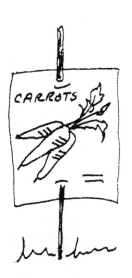

The garden's only response was the sound of plump tomatoes, too heavy for their stems, falling to the ground.

I could suddenly empathize with Dan Quayle, who had his own problems with another vegetable. I, too, have a new spelling for "tomato", but it wouldn't be fit to print.

I Go Under a Spell

I was hypnotized last week. My only regret is that the trance went by much too quickly.

The hypnotist gave a performance at the school where I teach. In much the same way early martyrs were chosen to fraternize with lions, I was chosen to be one of the subjects. Before I gave in to the flashing light and the lulling voice of the hypnotist, my last coherent thought was, "Please - don't let him turn me into a chicken".

Although I was spared that particular humiliation, I did manage to play an imaginary saxophone and I truly believed that my high heels, held up to my face, were a set of binoculars. The other stone-faced subjects and I rubbed imaginary suntan lotions onto our skin and we drank, as if there were no tomorrow, from imaginary glasses.

Being hypnotized was an amazing experience. While I was aware of the laughter and the applause, I simply didn't care. It was an out-of-the-body experience, not unlike coming out of general anesthesia or standing at the front door, waving good-bye to company who stayed for two weeks.

While our conscious minds were blissfully stuck in neutral, the hypnotist's voice overrode any anxieties. In fact, he told us something quite promising.

"Before you wake up, I'd like you to think about something you would like to change about yourself. Perhaps you would like to become a better student or a better athlete. Perhaps you would like to break a bad habit. Think about what you'd like to change while I count backwards from five. When I reach one, you will wake up. Five."

Although the idea of becoming a better student or

better athlete didn't tickle my fancy, I didn't miss the bit about dropping a bad habit or two.

In retrospect, I know that the hypnotist's grand plan was far from perfect. For example, a teacher who is three decades past her senior prom has had much more time to fine-tune bad habits than your average 15-year-old.

This is going to be easy, I thought. I'll give up junk foods and lose forty pounds this winter!

"Four."

Wait a minute. What am I thinking? What would life be like without potato chips? There must be something else. I know! I'll give up second helpings.

"Three."

What makes up a first helping? Isn't it true that some first helpings are rather small? Aren't they more like half helpings? Actually, a second helping can sometimes be like the second half of a first helping.

"Two."

I know! I'll give up coffee and diet colas. But then, how would I wake up in the morning?

"One!"

It was all over. I'm sure that if the session would have lasted three hours, rather than one, I would be a different person today. The next time someone offers to hypnotize me, I'll carry a list.

We're Sicker Than We Thought

Doctors have finally acknowledged a fact teenagers have know for 40 years - we're cool.

According to a study in JAMA, the journal of the American Medical Association, the "normal" human body temperature is 98.2 degrees, not 98.6 degrees.

If the study is valid, it's also true that most of us have been much sicker than we though we were during the past 124 years. A German investigator, Carl Wunderlich, set the old standard in 1868.

Being a teenager in the 1950s had its shortcomings - we were light years away from computers and floppy disks and most of us resembled oversized Barbie and Ken dolls. We were largely ignored when we made observations about body temperatures, such as "She's cool", "He's as cool as a cucumber" or "Cool, man, cool."

However, we persisted. Fabian was cool. Annette Funicello was cool. Dick Clark was cool. Our precocious observations fell on deaf ears.

It's hard to have much credibility when you're wearing a felt poodle skirt, bobby socks and bright pink lipstick.

Subsequent generations also failed to catch the attention of the medical community. The Beatles were cool. The Moody Blues were cool. Billy Joel was cool.

Had people listened to teenagers years ago, rather than to a doctor's study in 1992, our entire approach to sickness would have been different.

"Hi. I won't be coming into work today. I have a temperature of 98.6."

Countless hours of our lives have been wasted as we have waited for the thermometers in our mouths to reach the magic 98.6.

Imagine the number of times you have sat on one

of those paper table covers in a doctor's office with an alcohol-drenched thermometer stuck in your mouth.

"Hold it in your mouth a little longer," advises the nurse. "It's not even registering normal."

Four patients later, and when the thermometer feels permanently fused to the underside of your tongue, she finally rechecks the thermometer.

"Ah!" she exclaims triumphantly. "98.6. It's normal."

Wrong. By today's new standard, we are actually running low grade temperatures.

All of this confusion could have been avoided by listening to teenagers and, even the Beakniks, of the 1950s. If a future article in JAMA states that some people are literally "square", "far out" and "hip", my theory will be confirmed.

I Don't Do Crafts

With each approaching holiday season, I've suspected that I'm the only person around who doesn't knit, tat, crochet or turn bleach bottles into magnificent works of art. I don't make whimsical magnets for refrigerator doors, or Barbie clothes or tree ornaments.

The only object hanging on our refrigerator door is yours truly - and that only happens during television commercials.

According to the Hobby Industry Association, 77 percent of American households contain at least one craftperson, a 13 percent jump over last year.

In other words, 23 percent of us are buying their crafts and passing them off as our own. Although I may be the only one to admit this, I've been known to frequent craft bazaars and later give away my purchases as  "homemade gifts". If the giftee doesn't ask any questions, I don't bother to explain. Homemade is homemade.

My fraudulent nature also extends to bake sales. It didn't take me long to figure out that a pie, sold at a church bazaar, looks more authentic when served in one of my glass pie plates. An aluminum pie plate tends to raise suspicions. After transferring the pie from one plate to another, the illusion is completed by throwing handsful of flour on my face and the kitchen counter.

There are two reasons why I can't do crafts - my hands lack coordination and I'm missing what George Bush referred to as "the vision thing".

Some hands are specially designed to sew sequins onto tiny clothes for 11-inch dolls. Other hands can turn dish detergent bottles into Mr. and Mrs. Santa Claus.

It seems my hands were designed with certain limitations - they remove wrappers from candy bars and frozen pizzas. They also keep bracelets from falling off the ends of my arms.

As far as the vision thing is concerned, I simply can't look at a jumbled pile of jute rope and imagine a plant hanger. I can't imagine transforming a ball of yarn into a sweater. Putting an heirloom quilt together out of bits of cloth seems like an insurmountable task.

In spite of my shortcomings in the crafts department, I believe that I serve an important role in supply and demand. If the number of American households with craftpersons continues to grow, who's going to buy all of those things? Certainly not the people making the crafts.

Non-craftpersons, apparently a dying breed, also have a purpose in the great scheme of things. To paraphrase John Milton, "They also serve who only stand and buy."

Parents are Amateurs

Becoming a parent is very sloppy business.

While buying a house or car may require hours of paperwork, we run willy-nilly into parenting. Although we're given plenty of notice, we never consider drawing up contracts.

Given the hindsight of older parents, we now realize that contracts should have been a priority when our daughters were born.

Our contracts would have boldly stated, "I promise not to turn my parents into dumping grounds for troublesome phone calls."

They may live one or two time zones away, but our daughters feel compelled to reach out and touch us whenever they are faced with crises. Whether it's a car that won't start or a broken fingernail, we're the first to hear the news.

Quite amazingly, we're the last to hear how they fared once the crisis has passed.

Consider, if you will, these phone calls from the past few years:

"Dad? Someone's breaking into my apartment."

"Oh! I'm so glad you're home! I just thought you should know that my dorm's on fire."

"Mom? Terrible news! My name was lost on the school computer and I can't register for my classes!"

They somehow operate under the preconceived idea that we have all the answers to their problems. They seem unaware that any decisions, whether it be what to wear to work in the morning or which cereal to have for breakfast, are no easier for us than anyone else.

I had just fluffed up my bed pillow for the final time when the phone rang two nights ago. It was our daughter in Toronto, hardly a hop, skip and a jump away.

"Mom? Listen. Can you hear the sirens? My apartment building's on fire. I'm calling from the lobby. Do you hear the firemen shouting? Oh, I have to go now. They're sending us outside. Bye."

At least they feel better by calling us. That night we wandered aimlessly around the house, resembling zombies from a second-rate movie. As we chain-swallowed antacid tablets and wondered if our daughter had been turned into a human shish kabob, we waited for her return phone call.

Where would she sleep? Would she have clothes to wear to work in the morning? What about insurance? All of our questions went unanswered because the phone remained silent.

She called again last night and my first words were, "What about the fire?"

"Oh, that!" she answered. "It was just a small fire and they put it out right away. I was back in bed soon after I called. I hope you didn't lose any sleep over it."

No, of course not. By two in the morning, I had envisioned her phone melted down to a mass of dripping plastic. An hour later, I was imagining a homeless waif walking through the streets of Toronto, humming theme songs from "Annie" and "Oliver!"

No, we didn't worry at all.

Making Scents

Truth can be stranger than fiction.

An article in the New York Times reported that a Chicago smell researcher has been hired to invent a very special cologne. When sprayed on a car salesperson, he or she would smell - and this is no joke - honest.

It's very possible that the following comments might be heard on car sales lots within the next year:

"Let's buy the car, honey. Sure, it may have 140,000 miles on the odometer, but this guy really smells honest."

"Listen, you smell like an honest guy, so let's skip the games. Why don't I just pay the full sticker price?"

As you read this, sweet smells of success are being introduced across the country. According to the same article, smells, inducing people to buy, are being embedded in women's clothing, mixed with the air in stores and they're even being incorporated into automatic teller machines.

That's right. When dollar bills are treated with a mint fragrance, the money apparently seems fresher and newer. The subliminal effect on customers is amazing.

"Put the card in the machine again. Isn't this money wonderful?"

Unfortunately a dollar bill smelling like fresh mint has the same purchasing power as any other dollar bill.

Although subliminal scents are a rapidly emerging business, I experimented with the idea over 30 years ago, while I was still a freshman in high school.

I was the first to try vanilla behind the ears, and the effect on the opposite sex was amazing. I quite wittingly reminded my hapless victims of home baked cookies, home and Mom. It was a tough combination to beat.

To the utter dismay of my home economics teacher, I shared my classified weapon with my closest friends in home economics class.

As we splashed great quantities of vanilla behind our ears and on our inner wrists, our teacher was dumbfounded.

"Girls! Girls! This must cease! That bottle of vanilla has to last for the entire school year! Stop, I say, stop!"

However, when we mentally weighed the benefits of becoming femme fatales against having cooking vanilla for the second semester, seduction won out.

It was so simple. One day during band, Clyde, a very serious clarinetist who was later to become an insurance salesman, leaned over and asked, "What is that perfume you're wearing? It's sensational!"

"Um..Um..Um", I coyly responded. Because I was only a freshman, I was like the novice fisherman who could bait a hook but had no idea what to do once the fished was landed.

My trolling for "the big ones" ended shortly after another girl in the clarinet section began splashing oil of cloves behind her ears. Smelling like a gingersnap, she won Clyde by a nose.

Afraid of Being Overweight

Looking for another good excuse for weight gains during the holidays? Why not try fear?

Until I read the article about fish research being done in Sweden, my excuses for gaining weight during the winter months were getting fairly worn out.

"I just had a baby" was a perfect excuse for a couple of decades. However, a certain amount of credibility is lost when I'm forced to admit that our last delivery bill amounted to $500 and two chickens and our "baby" is now in her 20s.

It must have been a slow day at Sweden's Lund University when researchers decided to drop pike into carp-filled ponds. They must have run out of jokes to send over fax machines.

Out of fear of their drop-in visitors, the carp did what they had to do to survive their natural predators - they ate. In fact, after 12 weeks, they ate so much they were too big for the pike to swallow.

Once the pike are moved from the pond, it's believed that the carp will revert back to their slimmer shapes.

The same thing happens to me when I am weighed in a doctor's office. For years I've wondered why his scale shows a difference of 40 pounds over my scale at home.

The answer is simple. Because I'm so frightened by the time I've read 10 magazines in the waiting room and I'm finally called into the examining room, I mushroom to twice my normal size. The same thing happens when I try on bathing suits in a department store and the well-meaning salesclerk doesn't leave my side.

When I'm home alone, with no intimidating witnesses around, I'm a size three again. Honest!

That must be what happened to me during Thanksgiving. I was so certain that my dinner would be a complete disaster that I went up three sizes. I'm positive that eating four helpings of everything in sight was only a coincidence.

As soon as I leave for work in the morning, I'm surrounded by anxieties. From the fear that the neighbors' large maple will fall on my car as I back down the driveway, to the worry that a wall blackboard will crush me in the classroom, it's a wonder that I don't look much larger than I already do.

I may be a svelte size three as I leave the house, but by the time the last class bell rings at school, I resemble the Goodyear blimp.

I will no longer have to wince after the holiday feeding frenzy when people ask, "Carole? Is that you? Did you have some sort of fat transplant?" I will simply say, "I've been quite afraid."

If it works for carp, it should work for people.

Greasy Hair is "In"

For the first time in recent history, working parents are fashion trend setters.

A Los Angeles company recently introduced a product that gives clean hair the look and feel of hair that "hasn't been washed in three days."

I knew our time would come.

It's now possible for a single person to buy a tube of greasy gunk for their hair, guaranteed to make them look like a parent with no time for shampoos. Either that, or they look like they comb their hair with fried chicken.

Until now, I've been able to tell single co-workers from married-with-children co-workers by the way they look on the job. Single people always look like they've just stepped out of a shower.

However, if the greasy hair look catches on, we'll all look the same.

Although it's hard to believe that people would be willing to pay good money to look unshampooed and disheveled, perhaps they could benefit from other grooming habits I've acquired while raising a family. Balancing cooking, cleaning and washing clothes with a full-time career has resulted in some very interesting fashion statements.

The juggling act has involved standing up in front of a meeting and discovering that only one leg was shaved that morning.

It means rushing to work with a trail of toast crumbs clinging to the front of your sweater and a mouth faintly outlined with strawberry jam.

Balancing an outside job with a household schedule can involve grabbing your shoes as you leave the door and discovering, on the way to work, that one shoe is brown and the other is black.

While our daughters were all home, I rushed off to work one morning wearing one of their tiny pairs of pantyhose. By the time I had been tugging and pulling at my waistline for a couple of hours, I realized the error of my ways.

I hobbled around for the rest of the day with the gravitating waistband cutting off the circulation in my thighs and the crotch of the pantyhose at my knees. It wasn't a pretty picture.

Before the invention of non-static dryer sheets, I showed up for work several times with baby socks stuck to my clothing. It's hard to keep students' minds on their lessons when their teacher is shedding the family wash in the front of the room.

Instead of shelling out cash for products that make them look married, it might be easier for people to simply get married. The greasy-haired, disheveled look will follow.

The High Cost of Partying

No one does parties like Los Angeles.

According to the Los Angeles Times, city council members "approved a $175 service fee to be charged to the host any time police must respond a second time to quiet a loud party. An additional fee of $8 will be charged each minute a helicopter must be used in quieting the party." The vote passed 10-0.

Added to the cost of balloons and chip dips, that's quite a hefty tab for having a good time.

By L.A. standards, I must be running around with the wrong circle of friends. Parties in the Upper Midwest are mild by comparison, and they seem to be centered around two themes: buying items which will never be used and celebrating happy occasions, like weddings, graduations, new babies and birthdays.

The happy occasion parties are self-explanatory. Based around a universal menu (ham sandwiches, cream cheese mints and cake), they tend to be orderly affairs. They're traditions passed down from one generation to another.

Perhaps the most controversial parties are the ones known as "home demonstrations". Invitations are sent out to select lists of people - perhaps all relatives within a 1,000-mile radius or all of the names in the local phone book.

Cards are sent to the dubious guests, promising games, prizes and, of course, refreshments. It's an understood fact that, by the end of the evening, we will be introduced to the wonders of airtight refrigerator containers, wall hangings for every room in the house ("Turn your bathroom into the Louvre!") or Happy Hannah cosmetics.

Each invitation also carries a certain degree of obligation. In other words, "If you drink my coffee and eat my cake, you will buy enough cosmetics to last a lifetime or enough plastic containers to support an offshore oil rig."

As far as fun goes, I've never attended a "home demonstration" that has required a 911 call or a helicopter. There have been some hilarious moments - like the time the hostess was unable to make the plastic lid "burp", or the time I tried on plum-colored lipstick.

I also recall a wild evening when the game required me to drop clothespins from nose level into a plastic bottle. In spite of those moments of reckless mayhem, not once was a S.W.A.T. team called in to break up the party. Not once did a loudspeaker blare from a helicopter overhead, "Okay, ladies, drop those checkbooks and go home."

There have been times when I would have welcomed such an interruption. Compared to the money I've spent at home demonstrations, a $175 police service charge would be a small price to pay.

The Complexities of Buying a Car

In our household, buying a new car is a man thing.

As surely as swallows return to Capistrano and young men turn to thoughts of love, my husband begins making the rounds of car dealers at the first sign of spring.

By the time the first tree buds appear, he has already accumulated an impressive collection of slick brochures and exterior paint samples. In the same way that archeologists have searched for the Holy Grail, he is in quest of the ultimate car purchase.

To be perfectly honest, he gives more thought and study to selecting a car than he did in selecting a wife.

While he's in the major league of car buying, I'm only a rookie. As long as a car is red and has a tape deck, I'm satisfied. He's into horsepower, mileage and automotive terms I'll never understand.

He enjoys the ritual of buying a car. The price haggling and other negotiations are as natural to him as breathing. After accompanying him to a car dealer two years ago, I'm convinced that the mysterious system of purchasing a new car has been the downfall of the auto industry, rather than the economy. It's a tough game with complicated rules.

We were standing in the auto dealer's showroom, when I saw what I considered to be an ideal car. It was red. It had a tape deck. The chrome was shiny.

Sensing my enthusiasm - I recall doing cartwheels - my husband told me something most peculiar.

"No matter what happens, pretend you're not interested."

I acted disinterested long after the salesman escorted us to his tiny office. My disinterest became confusion soon after the game-playing began.

As I sat by silently, giving what I considered to be an Academy Award performance for apathy, my husband and the salesman exchanged preliminary insults, apparently part of the ritual.

"Let's see. Your car has 160,000 miles on it. I suppose we could junk it out and sell it for parts."

"Your new car in the showroom has a scratch above the driver's door."

After a few more friendly exchanges, the salesman wrote numbers down on a scrap of paper and pushed the paper across the desk for my husband to read.

It all seemed terribly secretive. There were only three of us in the small room, but the two men were writing clandestine notes to each other. The implication was that what they were writing couldn't be put into words.

"That's the best we can do," sighed the salesman.

My husband stared at the note for the longest time before he scratched out the numbers and wrote some numbers of his own. Once more, the paper was pushed across the desk.

"That's my top offer," said my husband.

Obviously it wasn't, because the scrap of paper made several more trips across the shiny desktop.

When it appeared that a settlement was close, my husband stood up and told the salesman, "We shouldn't be wasting your time."

My heart sank. The red car with the tape deck and shiny chrome wasn't going to be ours, after all.

I don't fully understand what happened then, but we landed up with the car.

Until buying a car is as easy as buying a box of breakfast cereal - when the price you see is the price you pay - I'll pass.

Competitive Weather

Storms are getting to be a dog-eat-dog business. Thanks to the media, regions are competing with each other for the most violent weather.

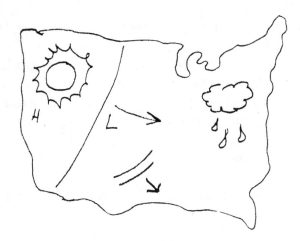

Not so long ago, it was hurricane Hugo, battering the eastern seaboard and putting hurricane Debbie to shame. Now the same thing has happened to Andrew, touted to be bigger and nastier than any other hurricane in our country's history.

Andrew was accompanied by a downpour of superlatives - largest, most devastating, most costly and most anything else that came to newscasters' minds.

"Andrew makes Hugo pale by comparison," noted one reporter on the scene. Others echoed his sentiments. Overnight, Andrew, the new kid on the block, had wimped out Hugo.

When I heard them, I wondered how those words affected the people who had experienced previous hurricanes. Were they thinking, "Oh, yeah, you might have known that a bigger and better one would come along."

We've become terribly competitive about weather. Imagine that you've received two inches of gully-gushing rain during a summer thunderstorm. You call your friend, not more than two miles away, to gleefully report your record rainfall.

"It sure was a downpour," he agrees. "We had four inches!"

Suddenly, you are no longer the winner. The score stands at 4-2.

Following a winter blizzard or a large rainfall, we tune into the weather report on the local television station. The inches of rain or snow are reported like sports scores by the excited weathermen, who must be normally bored with dewpoints and humidities.

"And it's Sioux Falls with ten, Worthington with five and ..."

We hang onto the edge of the couch in eager anticipation as he rambles through the names of towns. Finally he gives the name of our town and it's ... "Eighteen inches of snow!"

We feel a rush of adrenaline as the final scores are announced, not unlike having our number called in the national lottery. Victory is ours.

Knowing that we have had the strongest winds, the most hail or the most downed power poles brings out the hidden champion in all of us.

However, our Midwestern snowstorms are hardly noticed on national television. The reason is simple - we don't name them.

As with hurricanes, we know about most serious snowstorm days in advance. There would be plenty of time to pick out a name, call Dan Rather and make the most of it.

The names should be Nordic in nature - Eric, Thor, Leif, Ingrid, Ole or Gretchen - names that reflect austere, frigid climates.

An entire news team from a major network could be sent to a remote farmhouse, miles from the nearest town. During the height of the snowstorm, Rather or another news veteran could report what happens when power lines fall down, water lines freeze and toilets cease working. He could describe what it's like to be snowbound with a house full of children and a television that doesn't work.

After 24 restless hours of Uncle Wiggily and Chutes and Ladders, he would no doubt admit, "This has been the storm to end all storms. Ursula has made all other snowstorms pale by comparison."

If we would only be creative and name our blizzards, the rest would be history.

Avoid Cholesterol Deficiencies

Like high heels with pointed toes and tie-dye prints, the cholesterol controversy has come full circle.

Now that new studies are pointing out the dangers of low cholesterol levels, I'm glad I waited a while before jumping on the don't-eat-anything-that-tastes-good bandwagon. Extra low levels are now being linked to several unpleasant scenarios, including suicide and homicide.

I'm not surprised. I've seen the looks in some people's eyes at restaurants when I inhale a serving of greasy fries and they munch on lettuce salads. When I lean back in my chair, wiping telltale traces of black silk pie off my face with a napkin, I've seen their stares behind the mountains of salad greens.

For a long time I thought those looks were saying, "Having a nice time, honey? Don't you feel wonderful with all of those animal fats in your system? I'm having a great time too, sorting out these delectable carrot sticks and radish slices."

Now that the homicidal tendencies have been linked to low cholesterol levels, I realize those diners didn't have love and warm fuzzies on their minds.

Actually, they could have killed for a bite of my black silk pie. Only a few well-placed chairs and tables kept me from becoming a murder victim on the six o'clock news - "Cake-eating cutie bludgeoned by crazed tofu eater."

It's no secret that while cholesterol-conscious diners study menu selections with grim determination, the rest of us have been placing our orders with reckless, carefree abandon. It was only a matter of time before some scientist would discover the dark side of being too cautious.

Instead of offering diners a choice between smoking and non-smoking sections in their restaurants, it might be more prudent for the owners to offer patrons a choice between cholesterol and non-cholesterol sections.

While the fat-eaters would be offered conventional knives and forks, the others might be better off eating with utensils made of harmless wood or plastic. In the hands of the wrong person, a simple fork could easily become a lethal weapon.

Perhaps non-cholesterol diners could be sequestered in specially sound-proofed eating areas. That way, they would be less likely to be triggered into violence by the sounds of sizzling steaks. They wouldn't be forced to overhear comments, like, "Gosh, I'm stuffed!" or "If I eat another bite, I'll burst."

Less provocative words than those have sparked wars.

The new research suggests that anything can be overdone. Incredible as it may seem, it's possible to be too healthy.

Speeding in Minnesota

When it comes to speed limits in Minnesota, there are different rules for different people.

For example, the speed a car may safely travel seems to be directly related to the human population of any given area. You would assume that a car driving through a densely populated area would move at a snail's pace and that a car in a desolate, rural area would be able to fly with wind.

Wrong. The opposite is the case.

As my car flowed with the traffic between the southwest corner of the state and the Twin Cities last weekend, I found the car speed increasing from 55 to 70. By the time I reached the congested intersection of Interstates 35W and 494, I had become a veritable demon on wheels.

There were no possibilities of slowing down. Any attempt to comply with the posted speed limit would have resulted in certain death. The cars moved along as if glued together. As drivers passed, talking on cellular phones, adjusting their makeup and reading newspapers, I began doubting whether I would be able to make my exit.

The speedometer needle moved steadily upward...60...65...70. Surrounded on all sides by semitrucks, and car drivers flashing imaginative gestures on the other sides of their car windows, I was afraid I would be swept up in the great whirlpool of metal and glass and carried off to another part of the world. Perhaps Duluth or, heaven forbid, Cloquet.

As I white-knuckled my way past blurred billboards and too-fast-to-read highways signs, it dawned on me that travel in rural Minnesota is a far cry from travel in the urban areas. Even though the rush hour was at its peak, there wasn't a police car in sight.

You can travel along any highway in rural Minnesota with the guarantee that you will be fined for any decimal point over the speed limit. As your car crawls along from Point A to Point B, the only other life forms for miles around might be a herd of cows in a field or a couple of trees. Even though you might be driving the only privately owned vehicle on the road, you will most likely meet the cars driven by the county sheriff, a couple of his deputies and one or two highway patrolmen.

Because highway patrolmen are so prevalent in rural Minnesota and scarcer than natural blondes in the cities, we can only assume that two patrolmen are being assigned to each county in the state, regardless of population. Such a thing is possible in a state where legislators can debate for hours whether or not the polka should be the official state dance.

Urban car drivers no doubt find solace in the fact that 12 patrolmen have to watch two million drivers. They also don't live in fear of having their names published in the local paper if they're picked up for speeding.

There aren't enough trees to produce the paper for such a list in a large city. However, in rural areas, the list of speeders' names in the local newspaper is best-seller material. Taking up a prominent space on one of the inside pages, the speeders' names are printed under a heading called "Crime Log" or "Court News".

Maybe that's the difference. In rural areas, speeding is a criminal action. In the cities, it's a normal, everyday activity.

Camp Kick-a-Chunk

If my plan works out, the Upper Midwest will be renamed "Camp Kick-a-chunk" and thousands of California children will be heading here in planes, trains and buses.

A newspaper article reported that Southern Californians are paying thousands of dollars to have their lawns covered with man-made snow. North Hollywood Ice Co. has been grinding 300-pound blocks of ice into fine flakes that are blown onto customers' lawns.

Some folks will go to any extremes to have a "White Christmas".

Although the snow only lasts two days, the owner of the ice company noted, "Seeing the smiles on the kids' faces makes it worthwhile."

If watching snow melt on palm trees and diving boards is exciting for those kids, imagine how thrilled they will be when they're sent to "Camp Kick-a-chunk".

For the first time in their lives, campers will experience total immersion into the wonderful world of ice, sleet and snow. The camp will offer both crafts and survival skills.

Younger campers will be taught how to form perfect snowballs and snow angels. By the end of their three-month camp, they will be able to discern the differences between good-packing snow, yellow snow and "snert", the Upper Midwest's unique blend of snow and dirt.

Group activities for the novices will include kicking ice chunks from cars (hence the name of the camp), knocking icicles from house roofs and the game all of my children played at least once while they were growing up. For lack of a better name, the game could be called "Stick your tongue on a piece of cold metal and try calling for help."

Costs for restorative tongue surgery will of course be included in the camp fees.

Advanced campers will be taught how to weave their own parkas and they will have a working knowledge of the following terms: windchill factor, livestock advisory and whiteout. In their craft classes, they will create car emergency kits - coffee cans filled with matches, sterno heaters, candy bars and other essentials.

If parents are willing to part with $4,000 to have their lawns covered with artificial snow, imagine what they would be willing to pay to have their children sent to Camp Kick-a-chunk. Not only would those campers have a better appreciation of national weather maps on TV, but our part of the country could experience an economic boon beyond our wildest imagination.

To paraphrase the chant of the early miners to California, "There's gold in them thar chills!"

The Latest Health Threat: Pager Pudge

There are two more reasons why I'm glad I'm not a man.

Men's Health magazine reports that one of the medical maladies plaguing men in 1991 was "pager pudge". This condition affects overweight men who wear pagers on their belts. When they sit down, their spare tires flop over the pagers and muffle the beeping.

Although the article didn't go into the number of fire calls unheeded, business deals that fell by the wayside or surgeries that weren't performed, not responding to a pager sounds pretty serious.

The article also failed to mention another condition, "TV remote controlitis", which has taken over one particular member of our household. I won't mention names, but he has a mustache.

"Remote controlitis" has reached epidemic proportions. Harpers Index recently reported that 78 percent of men under the age of 30 are compulsive channel flippers during commercials. In the 60-and-over group, 55 percent of men are capable of watching 30 TV channels at the same time.

No reference was made to men in the 31-59 age group, but we can only presume that they were too busy changing channels to answer the phone survey.

The channel-changer in our house - the one with the mustache - exhibits classic side effects of this affliction. Because he can never be sure when a commercial will appear on the screen and he will be called into action, his neck muscles are like knotted ropes. Because there are so many buttons to push in so little time, he is rarely relaxed.

As he flips from one channel to another in rapid succession, anyone else watching television is lost in a hapless maze of broken dialogues and meaningless plots.

"Welcome, Barbara, to the White House living quarters. Here in the master bedroom there are two twin beds because...the Tar Heels have taken a 19-9 lead with only minutes left before halftime...Well, Mrs. Fletcher, I'd be mighty interested in knowing why you think the murderer is...Roseanne! Don't you have better things to do than stand around and...that's tonight's weather."

Night after night, the confusion continues. I never know how car chase scenes end or if the handsome, young detective lands up with the blonde. Watching television with someone out of control with a remote control is like putting together an impossible jigsaw puzzle.

After a few hours of channel switching, I invariably short-circuit. I get up from the couch and wander aimlessly about the house, mumbling monosyllables to the plants and trying to recall my name.

Things could be worse with the man with the mustache. At least his strenuous exercise with the remote control keeps him from having "pager pudge".

This is No Yolk

New Jersey's new egg law isn't all it's cracked up to be.

It is now illegal for the state's restaurants to serve eggs with runny yolks. Egged on by federal standards, calling for the destruction of salmonella bacteria by thorough cooking, state legislators passed the bill, which went into effect January 1, 1992.

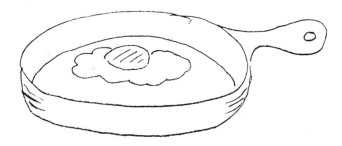

Needless to say, many of their constituents aren't happy about the law or the possibility of shelling out fines of up to $100. They think their legislators' priorities are scrambled. Like the eggs, the lawmakers are now in hot water.

Although some New Jerseyites may associate eating a half-cooked egg with the constitutionally guaranteed pursuit of happiness, it's not the first time that state legislators have landed up with egg on their faces.

As far as anyone knows, the New Jersey egg edict is the first of its kind in the country. In the not too far off future, other states may follow suit by banning rare steaks, cheese sauces or whatever else fits their fancy.

Carried to an extreme, laws governing how we eat our food may restrict us to government-issued cookbooks. Any Sylvester Stallone movie, showing the hero gulping down raw eggs, would be banned for national health reasons.

At the risk of mixing metaphors, any further legislation of this sort should help us determine which politicians are bad apples or good eggs.

Personally, I like to have my eggs totally mutilated and as hard as rocks before they are eaten. The end result, which looks like something found in a cow pasture, isn't a pretty picture, but the idea of eating a dripping egg isn't very appealing either.

My husband prefers to have his eggs done over-easy. In other words, if you would listen carefully to the eggs on his plate, you might be able to hear some faint chirping.

As long as he doesn't make me eat eggs his way, I make allowances for the traces of egg which cling like epoxy glue to his plate and fork.

Although I prefer to know that an egg is thoroughly dead before I eat it, I respect the wishes of people who like to eat their eggs with gelatinous whites and runny yolks.

In terms of individual freedoms, the New Jersey legislature has definitely laid an egg.

Men's Calf Implants

According to an article in Men's Health magazine, men are no longer content with hair plug transplants, capped teeth and overdeveloped biceps. They are now turning to calf implants.

During the $4,000 operation, plastic sheaths are inserted into their calves. The procedure will hopefully transform the patients into Arnold Schwarzenegger look-a-likes. Depending on the thickness of the plastic in their bulging calves, some men might appear to be smuggling watermelons in their socks.

It's unfortunate that men haven't learned a lesson from women with their silicone implants. Las Vegas chorines come to mind. For many years these disproportionate women have been the targets of lewd jokes because they decided to live better through chemistry and improve upon nature.

Their legs may look like stacked-up bowling balls, but are men ready for their co-workers' reactions? Will their cumbersome calves be perceived as open invitations for harassment?

"Sure he has nice legs, but can he type?"

"He only got the job because he showed a little leg."

On tennis courts, on gold courses and at swimming pools, overly endowed calves might give men a sense of what women have tolerated since the beginning of time.

"Look over there, Janine. Nice legs!"

"Yeah. But are they real?"

"Hey, Janine, get a load of those gams. Does he look easy or what?"

If men decide to have their calves enlarged, it's very possible that women may be led to say to each other, "I don't know about you, but I'm definitely a leg woman."

The calf implants may also confuse archeologists 2,000 years from now, who will discover bones lying side-by-side with non-biodegradable plastic cushions.

"What do you make of these, professor?"

"I have no idea, but it's clear they go all the way back to the Plastic Age. Perhaps the pads were part of some ancient male ritual."

In order to avoid unpleasantness in the present and dark mysteries in the future, men would be much better off enlarging their calves through simple exercise.

Although physical exercise has never been one of my virtues, my calves became ample and well-rounded after years of standing on cement basement floors while folding the laundry. Climbing the stairs with armfuls of laundry also helps.

Any male who wishes to have a leg up on this enlarged calf business is encouraged to do the same.

The Case for Chocolate

What better time is there to talk about chocolates than Valentine's Day?

Although I don't need a reason to eat chocolate, a New York psychiatrist has been doing some lengthy research into the subject. Dr. Michael Liebowitz has found that romantic breakups often lead to longings for chocolate. He theorizes that romantic feelings produce phenylethylamine, a substance also found in chocolate.

I don't want to rain on Dr. Liebowitz's parade, but my romantic life is quite pleasant, thank you. As far as research goes, the good doctor failed to observe that certain individuals have chocolate deficiencies. As for phenylethylamine, how could you possible blame something you can't spell?

In my case, my chocolate deficiency is linked closely with my cholesterol deficiency.

Fats are vastly underrated by the medical community. It's my belief that fats are food for the hair, skin and one's general good health.

Hair products are a multi-million dollar business in this country, but a few decades ago people didn't bother with hair conditioners. After a weekly shampoo, their hair continued to be lustrous and shiny.

If you doubt this, look at your average farm dog, who exists not on some high-tech dry cereal based on his age, but rather on table scraps. If that dog has a sufficient daily intake of meat trimmings and pan drippings, he has a healthy, shiny coat of fur.

If people would eat more fats, they could save a fortune in hair products, which are designed to make you look as if you eat fats by the gallon without any of the fun.

Instead of smothering our skin in emollients and

moisturizers, we could much better eat an order of french fries. French fry oils seem specially designed to go straight to our faces.

Fats and oils are synonymous words, and who would deny that oils are necessary if we hope to keep machinery in tip-top condition? In the same sense that we wouldn't run a car without oil, doesn't it seem that our bodies, in general, deserve equal consideration and attention?

In the same way that a car's engine and transmission have to be well-lubricated, so do our vital organs. If this intense anti-fat campaign continues, it's very possible that we may see cases of kidneys rubbing against spines or lungs punctured unnecessarily by nearby ribs.

Personally, I would hate to have my cause of death listed as a blown engine, or as my friendly car mechanic refers to it, "meltdown".

For all of these reasons and because chocolates also happen to be some of my best friends, I will not forsake them. I will continue to eat chocolates every five miles or every 30 minutes - whichever comes first.

Generic Gym Shoes

Although the news comes too late for our family, I'll share it, anyway.

According to New York City market researcher Irma Zandl, teenagers have lost their fascination with fancy athletic shoes. You know the kind of shoes I'm talking about - the leather ones endorsed by professional basketball players and featuring pumps, jelly-filled soles and prices comparable to the national debt.

For too many years parents have been taking out second mortgages on their homes and working extra hours so their kids can wear the special shoes, which all land up smelling like last week's fish after exposure to the elements.

The new "with it" footwear is low-tech tennis shoes, including the high-top canvas shoes which were worn during the 1950s and '60s.

Parents can only hope the same turn-around will happen with designer blue jeans.

When our oldest daughter set out for college in 1984, she packed several, new pairs of designer jeans. Although generic blue jeans had been fine during high school, she rolled her eyes and exclaimed, "I simply have to have them. I don't want to look like a nerd at college."

On her first visit home, after a short three months, she wore a pair of jeans I had never seen before. Held together by a wish and a prayer, the jeans' rips and tears revealed large expanses of insulated underwear.

I clutched at my heart, a nervous habit passed down by generations of women in my family.

"But what happened to the Lizzes and the Guesses?" I finally gasped.

"I traded them for these! Aren't they absolutely wonderful?"

124

Perhaps other people's children will find this same happiness with canvas tennis shoes.

If this designer label craze runs full circle, parents will have to find other things to talk about with their offspring. They will no longer have to say, "We're sending you to school, not a fashion show" or "Two hundred dollars for a pair of shoes? Do you know what I wore for shoes when I was your age?"

Gone will be the stories of how we wore the same pair of Buster Browns for 10 years, or how our fathers used to nail old pieces of car tires onto the bottoms of our shoes when the soles wore out. Tales of the Great Depression may be finally laid to rest.

No parent will have to exclaim, "I don't care how many of your friends have pump athletic shoes. If they jump off a bridge, will you too?"

When their children begin wearing sensible (translation: $5 a pair) tennis shoes, few parents will have reason to protest, "We're not made of money, you know."

Life will become much less complicated with canvas, no-frills tennis shoes.

My Good Friend, Doris

My husband normally doesn't criticize my choice of friends. Perhaps because love is blind, he usually accepts my friends, regardless of shape, size or persuasion.

His attitude changed drastically this winter, when I became friends with Doris, who sells cosmetics at a large department store in a nearby shopping mall.

Doris and I became instant friends when we met. She liked my skin tone, so I bought a small fortune's worth of skin creams. She like the contours of my mouth, so I bought several tubes of lipstick. The more she admired my various attributes, the more I bought.

Doris often sends cards and letters to our house, giving me inside information about preferred customer sales and other sale events. In her letters she encloses little, handwritten notes - "I've missed seeing you" or "Please look me up during your next trip to the mall."

It's nice to know that Doris, day in and day out, is waiting breathlessly for my return visit to her counter.

Quite out of character my husband told me, "You know, buying a tube of lipstick is hardly the basis for a long-lasting relationship."

For some unknown reason, he becomes uncomfortable when salespersons in expensive department stores greet me by my first name. He simply doesn't "get it." I suspect he would be much happier if I would walk up to a store counter and say, "I would like to purchase your most inexpensive cologne. Yes, Eau de Barn would be nice. Here's your two dollars."

He doesn't seem to understand the psychology - or the price - of maintaining one's beauty.

When Eileen in lingerie, who thinks I look much younger than 49, or Betty in jewelry, who thinks gold is definitely my color, greet me in the store like long-lost friends, my husband seems unusually nervous. Matters become further complicated when my retail friendships go beyond first names. When I discover that a salesperson is putting five children through college, or another is trying to put together her life after losing everything in a housefire, I tend to buy more than I need. I feel compelled to purchase enough items to keep them on their feet.

When it comes to friendships, there are no credit limits.

Doris is a persistent friend. She recently began making phone calls during the evenings. She apparently misses seeing me.

If my schedule slows down, I might drive over to the mall and surprise her. It would really be a friendly gesture.

Why My Husband Won't Sign a Living Will

Forget the old adage, "Where there's a will, there's a way." Writing a will, particularly a living will, can lead to some revealing discussions.

While signing copies of our wills last week, my husband and I had no problem in confirming "What's yours is mine and what's mine is yours." Actually, it's a much better deal than my current policy regarding our dual income: "What's yours is mine and what's mine is mine."

When your most valuable possessions include a miter box, cookbooks in mint condition and several junk drawers in the kitchen, there is little basis for contesting a will.

Even though my husband questioned whether I should sign a document stating that I was in sound mind, the will-signing was fairly uneventful. That is, until we began discussing living wills, which turn over our eventual health care decisions to other people.

As we were about to sign this second document, my husband leaned back and observed, "I'm not sure I would like to have a person make those decisions who would directly benefit from my dying."

His statement caught me off guard. It seemed as though he was expecting to have a decision-maker chosen at random from an out-of-town phonebook. Did he truly expect a stranger from Kankakee, IL, or Sheboygan, WI, to drop everything at a moment's notice to tend to his needs?

"Trust me," I coaxed to no avail. I suppose he was considering my other decisions in life. Like the time I removed a large, dead branch from a tree and broke

128

three living room windows in the process. Or the time I backed up the car and tore off the sideview mirror. Perhaps it was the time I failed to read the fine print on the record club ad.

Sound decisions haven't always been my main strength.

"It's your lists", he finally admitted. "I can see it now: clean out refrigerator, strip floor wax, unplug husband."

In a way, he spoke the truth. I am committed to writing lists and I do enjoy crossing out each completed task with a certain reckless abandon - almost willy-nilly, if one will pardon the expression.

"You're wrong", I blurted out. "If I did have such a list, you would certainly be given top priority. You're much more important to me than the refrigerator or the floor wax."

The battle of the wills remains unresolved. The more I try to convince my husband of my good judgement, the more I sound like Lizzie Borden. I only hope that the stranger in Kankakee doesn't write lists.

The Haggis Ban

While Congressional perks have become accepted practices, there are still certain things our government can't stomach.

In January the U.S. took a decidedly strong stand against haggis - considered by many to be the national dish of Scotland. Die-hard fans of Robert Burns had hoped to import the delicacy into this country to celebrate the poet's birthday.

Just in case you're one of many who can't tell a bagpipe from an accordion or you have been led to believe that Robert Burns is the name of a well-known cigar, there's something you should know about haggis.

It's not a pretty picture. The dish contains sheep's heart, lungs and liver mixed with oatmeal, onions and black pepper - all boiled in a sheep's stomach and served with mashed turnips.

Because U.S. officials declared the delicacy unfit for human consumption, I'm only grateful they haven't decided to check out my kitchen. It has seen its share of questionable menus and recipes run amuck.

I created one of my most forgettable meals shortly after we were married. The waffles, stuffed green olives and grape soft drinks were colorful on our new china, but my husband must have regretted not putting an escape clause in our wedding vows. The meal gave new meaning to the words, "For better or for worse."

Although no one has died from my cooking and my picture doesn't appear on post office walls, certain experiments with new recipes put me right up there with Typhoid Mary. My souffles have refused to rise to the occasion and I have had dismal experiences with never-fail fudge. One time I even managed to burn a hard-boiled egg.

My cooking fiascoes can be attributed to two factors: I rarely measure and I substitute ingredients with reckless abandon. For example, if a recipe calls for some exotic ingredients, such as Mediterranean capers, I simply substitute with something else of the same color group - perhaps green beans.

By substituting ingredients according to color I go one step beyond the wimpy suggestions listed in most cookbooks. Under my system, miniature marshmallows may be substituted for potatoes and finely chopped beef may be replaced by chocolate chips.

In our house, each meal is an adventure - a journey into the unknown.

Once in a great while, the substitutions work and someone will exclaim, "This is wonderful! You must give me the recipe!"

Instead of confessing that the dish is the result of whatever happens to be in the house at the time, I murmur, "Sorry. It's a secret family recipe."

I don't have the heart to tell them the truth. Perhaps the Scots should have done the same with their haggis.

The Michigan Fungus

A popular expression from my youth - "There's a fungus among us" - has become a reality.

Scientists have discovered the largest and oldest organism known to Earth on a forest floor in Michigan. Dating back 10,000 years, the fungus covers 37 acres and weighs at least 100 tons. Because no one seems to know the lifespan of a fungus, it could still be a kid.

In everyday terms, the fungus would produce more cream of mushroom soup than most of us can imagine.

The age of the fungus may be an undisputed fact among scientists, but I have my own theory about the growth's origin. If scientists would check the area more carefully, they would most likely find an abandoned refrigerator and a plastic leftovers dish.

Over the years, our refrigerator has become a spawning ground for various life forms, including furry fungi. Although a 100-ton fungus would have sent the refrigerator crashing into the basement long ago, the life forms on the shelves have ruined many otherwise healthy appetites.

It wouldn't be surprising to hear someday that another casual homemaker in Michigan was the cause of the gigantic growth.

Because I'm from Michigan, I know that the residents of the northern peninsula have a certain resiliency and that they will be able to capitalize on the fun part of fungus. You have to be creative when the rest of the state looks like a large hand and you live on a hunk of ground resembling a poorly-cut piece of pizza.

If nothing else, the still-growing organism should provide plenty of excuses for residents.

"The fungus ate my homework."

"I'm sorry to be late for work, but the fungus was blocking the road."

Any fungus of that size probably has a personality of its own, and it probably does whatever it pleases.

Contests could be formed to name the incredible hulk from the north woods. "Fergie", "Gus" and "Fred" would be likely suggestions.

Taking a cue from Kansas City, an entire theme park, Worlds of Fungus, could be built in the area.

They might also take a crack at being billed as the Toadstool Capitol of the World.

Any of these possibilities might work if the fungus watches its weight and stays under 100 acres. If not, the only claim to fame for residents might be a Hollywood movie, "The Fungus that ate the upper Peninsula."

A giant fungus. What a great gimmick. Any way you look at it, there's gold in them thar woods.

Minnesota's Official Dances

The truth is out. During its last session, the Minnesota legislature failed to designate the state's official folk dance.

Although it's small comfort, Minnesotans should sleep better at night, knowing they already have an official state muffin (blueberry) and an official state drink (milk).

The dance controversy sparked hours of heated debate about the relative merits of the polka and square dancing.

Before the legislature reconvenes, I would like to nominate the following fancy footwork, common not only to Minnesota but also surrounding states:

The Shopping Cart Shuffle - This dance is performed while standing in line at the grocery store. As ice-cream melts in your cart and the person ahead of you recatalogues a foot-high stack of cents-off coupons or scratches a lottery ticket with a coin, the dance pace intensifies. By the time the ice-cream is dripping onto the floor, the Shopping Car Shuffle has been known to lead people over the edge.

The Jumper-Cable Two-Step - In order to get into the right mood for this dance, you must be standing along a deserted road in subzero temperatures with a dead car battery at your side.

The Bathing Suit Boogie - This dance is achieved by trying to squeeze this year's size 14 figure into last year's size 10 bathing suit. Accompanied by grunts, groans and heavy breathing, this dance should not be attempted in public or in the presence of small children.

The Molar Mambo - Accomplished by sitting in a dentist's waiting room while listening to drilling noises in the next room, this dance is usually accompanied by balancing an outdated magazine on one's lap.

The Bullhead Bounce - Best performed on the end of a dock, these intricate steps are directly related to how the fish are biting and the proximity of the nearest rest room.

The Remaining Days Rumba - This dance reflects the anxiety in a classroom during the final days of a school year. It involves considerable clock watching, fidgeting in a desk and having a vacant look on one's face. Actually, that's how the teachers feel - I can't speak for the students.

Some critics say that determining an official state dance for Minnesota is a sheer waste of time and taxpayers' money. On the other hand, supporters say such rulings help people feel part of the legislative process.

Terence, a Roman writer who never lived in Minnesota, once noted, "They who love dancing too much seem to have more brains in their feet than in their heads."

He might have felt the same way about certain state legislators.

Our Marriage is Par
for the Course

We had just finished breakfast one morning this week when my husband gave me one of his come-hither glances.

I've become familiar with that particular glance since our children grew up and left home to pay their own bills. It's a glance promising sweaty palm excitement and spine-tingling thrills.

The glance says - "Let's play golf".

By itself, the expression "playing golf" is an oxymoron. Playing should be a pleasurable experience, not a lesson in humiliation. We can't play golf any more than we can "play war" or "play tooth extraction".

If it's a game, an opportunity to enjoy the great outdoors, why do we bother to keep scores?

When we add a spouse to the game, it's a whole new set of dynamics.

Golfing with my husband has become a bittersweet experience. I want his game to go well for him but, at the same time, I would like him to experience the agony of defeat once in a while.

From my side of the fairway, it's easy to forget the appliances he's fixed around the house and the breakfasts he makes on Sunday mornings. For two hours, as we meander from one green to the next, he becomes my antithesis and the focus of my competitive nature.

It would have been much better if our wedding vows would have mentioned, "To love, honor and respect each other on the golf course."

While unloading his clubs from the car trunk, my husband undergoes a Dr. Jekyll/Mr. Palmer transformation. By the time the cleats on his feet, he has forgotten that I am the woman he chose out of thousands and that I'm the mother of his children.

If it has taken me four strokes to reach the green and his ball is lying inches from the cup in only two, there's a certain Chestershire Cat grin on his face that only a nine-iron could remove.

"So," he will ask in somber Machiavellian tones, "what are you lying?" He asks, although he already knows.

"Four," I murmur.

"I'm here in two", he responds, assuming, of course, that I can't count.

And so it continues through the entire golf game.

Once the golf clubs have been packed into the car trunk, things return to normal. I can even imagine living with him for the rest of my life.

Some marriages are partnerships. Others are based on friendship and mutual interests. While all of those ideas are wonderful, I would recommend the game of golf for added excitement.

The Cooking Classes

When I turned the big five-oh a few weeks ago, I decided to turn over a new leaf. In my case, it was a bay leaf.

I enrolled in culinary classes and I've learned that "gourmet" doesn't include casseroles made with cream of mushroom soup. I've also discovered that pans received as wedding gifts can be used more than once.

If nothing else, the classes have made our marriage spicier. It was also interesting to find out that sage, rosemary and thyme are more than song lyrics made up by Simon and Garfunkel.

Instead of thawing out frozen pizzas, I'm now into making vinaigrettes - exotic combinations of vinegar, oils and herbs. In fact, during the week following my introduction to vinaigrettes and a smattering of French expressions, that's all I prepared.

By the end of the third day, my husband poked at his salad and asked, "Is there life after oil and vinegar?"

"C'est la vie", I purred, giving what I considered to be my best imitation of Brigitte Bardot.

During that first week, we consumed so much olive oil that we began to think of Italy as the old country. I've purchased enough oil to coat all of California's sunbathers - enough to create a decent-sized oil slick.

We went through so much vinegar that our lips were drawn into permanent puckers.

Actually, that's not as bad as it sounds. When you walk around all day with puckered lips, looking as though you expect to be kissed at any moment, married life takes on an exciting, new dimension.

With so many gallons of vinegar in our system, we should be well preserved, if not pickled, until the next glacial period.

My classmates and I have been encouraged to rely on taste rather than measurements in our culinary efforts. That's fine with me because I rarely measure - it's much too scientific. It's a known fact that the scientific part of my brain isn't user friendly.

Like my mother and her mother before her, I've relied heavily on pinches, dashes, smidgens and glops. It's also the reason I've never prepared the same dish twice.

By the end of our second class, the chef told us, "Be creative. Use your imagination."

It was dangerous advice.

With almost missionary zeal I began using basil in everything. All of my meals began taking on an eerie, greenish tint. Although my husband tolerated dandelion greens in his salads, he did impose a botanical ban on pansy and nasturtium blossoms.

I'm also experimenting with a new vocabulary. My friends and loved ones are no doubt impressed when I say "chiffonade" instead of "chop" and "beurre blanc" instead of "white sauce".

They must be impressed. Why else would they be speechless?

Inflatable Bikinis

Clothing designers have a bad sense of timing.

Now that I no longer appear in a bathing suit with witnesses around, they have solved the pyramid-figure problem.

A bathing suit manufacturer now offers an inflatable bikini with a plastic pump that air-tanks the top. Taking a cue from the pump athletic shoes which promised more bounce, they hope to do much the same from women who resemble ironing boards or their 12-year-old sons.

If the fad catches on, we should expect to find pump shorts for the derriere-less and pump shirts for men too busy to work out in a gym. We should also expect unusual results from clothing that expands upon demand.

"I know this dress looks tight, but I overinflated this morning."

"Where's the nearest service station? My bathing suit has developed a flat."

The inflatable bikini tops might prove beneficial for non-swimmers who feel they're too old for flotation devices decorated with likenesses of Bart Simpson or the Care Bears.

Although policemen and firemen have worn lifesaving clothing for years, inflatable bikini tops might be the first example of lifesaving clothing for swimmers.

Now that women are able to pump their way to dubious popularity on our beaches, clothing manufacturers will undoubtedly seek other ways to capitalize on women's basic fears of inadequacy. Radial bikini tops might be one possibility, or consider the benefits of bikini tops with non-slipping snow treads.

When the pumps fail to function, should they be

returned to the clothing store or do we simply take the bikini tops to a tire shop to be patched or re-treaded?

When women, pardon the expression, tire of pumping up to an additional one or two inches, what's going to keep them from turning to heavy-duty, commercial air compressors? Will air pumps at gas stations be restricted to females over the age of 18?

If a woman should accidently overinflate her bathing suit, could she possibly become airborne as a polka-dotted or flower-printed dirigible?

Whatever the outcome, inflatable bikini tops certainly give new meaning to the expression, "pumping up one's confidence."

Now, if bathing suit manufacturers would come up with a button to deflate hip measurements, I might be interested.

Pass the Hormones, Please

Hormones are fickle, changeable little things. They may be here today - and gone tomorrow.

For the first half of my life, the school, my parents and the church told me I had too many hormones. If I should live to be 100, I'm now entering the second half of my life, and it's the general consensus that I don't have enough.

My high school physical education/health teacher was the first to broach the subject during the 1950s. Her health class lectures didn't go beyond, "Look at the chart, girls. This is your liver and this is your heart", but she did manage to tackle the topic of our raging hormones in a manner we all understood.

She showed us 10-minute black-and-white movies with titles like "Nice Girls Say No" and "You're a Young Lady Now". The movie plots left something to be desired, but the message was clear: "You are now entering dangerous territory."

If the girls' discussions in the locker room would turn to racy topics, like Frankie Avalon or Pat Boone, our teacher would appear out of nowhere, bellowing her favorite words of advice: "Take a cold shower!"

To our 15-year-old minds, a future filled with cold showers didn't seem all that promising.

In church classes the story of Adam and Eve was repeated annually for our benefit, but the hormone part of the story was skimmed over in the blink of an eye. I did learn, however, never to accept apples on a date and that another word for hormone was "original sin".

For the next few decades, I peacefully co-existed with my hormones.

It wasn't until recently that my behavior necessitated a second look at them. When my body's thermostat fell out of sync with the rest of the world and I began fantasizing about living on an iceberg, I decided to seek medical advice.

"I think I need hormones", I told the doctor.

"Hmmm," he replied, in a way that only a person with ten years of post-graduate studies can respond. "Anything else?"

"Well," I added helpfully, "I'm having these hot flash kind of things, I've begun looking at waitresses and I've lost all interest in lingerie catalogs. Does my voice sound lower? I think I'm becoming androgynous. Does it seem warm in this room? Am I talking too much?"

It was more than he wanted to know. I am now living better through chemistry. I no longer turn on the air conditioner during the winter and I no longer preface every conversation with "Whew! It's hot!"

It is disconcerting to know that, with all of this "change" business, I've become my mother. The only difference is that she talked about the topic in hushed tones at the kitchen table with her friends, while their husbands watched TV in the living room. I've gone public.

We've come a long way.

Behind Closed Doors
with Amoebas

It's funny how one small incident can change the direction of one's life.

Whether it's standing under the wrong tree during a thunderstorm or picking winning numbers in a lottery, one decision can change the course of our lives forever.

In my case, amoebas had a large influence on what I became as an adult. Had I not given the wrong answer on a final exam in Zoology 101, there's no doubt in my mind that I would have become a world-renowned scientist or at least a name on a Trivia Pursuit card.

I remember the question well: "Amoebas are (a) asexual (b) bisexual or (c) sexual." Although I couldn't imagine much social life for a creature with one cell and a nucleus, the amoeba did have more appeal than some of my dates in college, so I circled "c".

Discouraged by the red-ink slash mark across my answer when the exam was returned, I began concentrating on my English literature classes, where creative answers were the norm, rather than the exception.

According to my professor, who apparently was the happiest when he could be red-inking tests or discussing mitotic division, amoebas don't go through dating and marriage rituals. They don't experience kisses at the door or giving bouquets of flowers. They simply split in two when the time seems right.

A scientific report released recently suggests that amoebas aren't dull, but rather discreet. French geneticists have examined them carefully and have discovered telltale signs of "sexual mixing." Evidently, that's scientific jargon for "fooling around."

Just because amoebas haven't revealed their innermost secrets on daytime talk shows or bragged with other amoebas in locker rooms doesn't mean they don't have fun. Actually, we should applaud their sense of propriety during a time when no subject appears to be taboo on television.

I can imagine now what the amoebas talked about when they were placed on glass slides in our zoology labs.

"What's the matter, Henry? All of a sudden, you're so distant."

"Good grief, Harriet, they're watching us. That's not the sun up there - it's an eyeball."

Although the news about amoebas comes 30 years too late to affect my grade point average, it's consoling to know that my answer was actually correct.

Now, if someone could prove that author George Eliot was really a man or that the shortest distance between two points is a crooked line, I would have all the makings of a Rhodes Scholar.

The Loss of a Diamond

One out of three isn't all bad.

During my past 29 years of married life, I have lived in fear of losing my figure, my engagement diamond and my teeth. As of this week I still have my teeth.

Shortly before Valentine's Day, I had just finished washing the dishes when I glanced down at my ring. Instead of a shining diamond, I could only see the gaping hole left in the setting.

A friend took the sink drain apart and the two of us resembled gold-panning prospectors as we closely examined the sludge left by decades of dishwashing. Although hopes were high as we extracted shiny bits of tin foil from the drain, there were no shouts of "Eureka!"

Floors were swept, car seats were overturned and every corner of the house was examined on hands and knees. Because I couldn't be sure when I had last seen the stone, it was like looking for the proverbial needle in a haystack.

I was devastated.

It may not have been the Hope Diamond, but it had much sentimental value. Over the years it symbolized the resiliency of our marriage and I had literally worn it through thick and thin. At least 20 diets, to be precise.

I wore the diamond when I prepared one of my most memorable meals as a new bride at the University of Iowa: grape Kool-Aid, stuffed green olives and waffles without syrup.

As the years passed, the ring became part of my left hand. Whether I was in a hospital labor room, or at a grade school program or standing in line at the grocery store, I knew that if I glanced down at my hand, the ring would be there.

Somewhere along the way, I assumed that the ring and all that it came to represent was my moral responsibility. Over the years, it had become too valuable. I had the prongs on the ring checked often by jewelers, and they did their best to assuage my monumental fear of LOSING THE DIAMOND.

When I called my husband to tell him about the lost diamond, he helped put everything into perspective.

"Don't be so upset," he told me. "It was only a thing. We still have each other and that's all that really matters."

I don't know what kind of reaction I had expected from him, but he said all of the right words - just in time for Valentine's Day. As he spoke, a huge burden was lifted from my shoulders.

Perhaps the diamond will be replaced someday, but it will never be the same. Given a choice, I would much prefer another 32 years with the same husband.

Doing the Laundry

She had to be joking.

Actress Whoopi Goldberg told Woman's Day magazine, "I love doing laundry. I even make it up; I mean, I throw things on the floor to make laundry. That's how much I love it."

Either she was joking or she has taken too many accidental falls down the clothes chute.

Being confined to the basement for hours and turning dirty socks right-side-out is not my idea of a good time. In fact, most of my adult life has been spent creating ways to avoid doing the laundry.

While they were growing up, our daughters saw the clothes chute as an alternative to hanging clothes up and wearing them again. The clothes chute was the perfect "out of sight, our of mind" solution.

Convinced that I had better things to do with my time than wash, dry and fold, I tried my best to have clothing items and bath towels used more than once before they were dispatched to the laundry godmother in the basement.

My first inspiration came from their collections of scratch-and-sniff stickers. If they were willing to trade their sack lunches at school for a handful of the paper stickers, why couldn't the same importance be attached to a T-shirt with spots that smelled of last night's spaghetti?

"Take a whiff of this sleeve," they could tell their friends. "That's where I leaned too close to the pizza last Saturday night."

When that plan failed, I tried visual effects.

One day I gathered them together in the kitchen, and waved a damp bath towel in front of their faces.

"See this towel? How dirty can it be after someone

has taken a two-hour shower? It could be hung up and used again!"

They looked at me in disbelief as I continued my sermonette, including small vignettes from what they fondly referred to as "the old days". Sensing I was on a roll, I told them how bath towels were often shared with other members of the family and were only washed once a week.

"Gro-o-ss!" my daughters cried out in unison. It was one of their favorite responses. They were so good at it that the word was usually recited in two-part harmony.

Although my logic fell upon deaf ears that day, things did change in Operation Laundry. Unbeknown to the rest of the household, the daily mountain of damp bath towels was simply fluffed in the dryer and refolded.

Everyone was happy. The girls loved the smell of clean towels, fresh from the dryer, and the laundry godmother spent less time in the basement.

Junk Phone Calls

If junk mail can be stopped from coming to our homes, why can't the same be done with junk phone calls?

I don't believe unsolicited phone calls were what Alexander Graham Bell had in mind when he invented the telephone in 1876. According to telephonic legend, Bell accidentally spilled some battery acid on his clothing and yelled to his companion in the next room, "Mr. Watson, come here, I want you!"

He didn't ask, "Mr. Watchtonson, could I interest you in some aluminum siding?"

That's one of the interesting things about unsolicited phone calls. The caller invariably mispronounces your name. A simple name like Smith can land up having three syllables, and it's a sure sign that the caller doesn't know you from Adam.

According to the Harper's Index, an estimated 200 unsolicited phone calls are made by U.S. telemarketers each second. That's letting your fingers do the walking 12,000 times a minute or 720,000 times an hour. Most of the calls are made during the evening, when the average American has better things to do than trip over furniture and the family dog while trying to answer the phone.

Until some future court determines that junk phone calls are an invasion of privacy, there are a few things we can do to undermine the "let's-call-a-million-homes-tonight" mentality.

Depress the caller. Slamming down the receiver or saying "No" only encourages them to call you again when you're in a better mood. Instead, let the caller think you were the account that got away.

"Would I be interested in aluminum siding? Gosh, I wish you would have called last month. We had a call then, and we have spent over $600,000 on new siding."

"A new vacuum cleaner? If only you would have called yesterday! Our old vacuum broke down and we were desperate for a new one. We went right out and bought one - it was the top of the line, too. Money was no object."

Given enough of those responses in one night, any caller in his or her right mind would seek another line of work.

Use up their dimes. Have you ever noticed how the callers don't ask for a response until they have gone through a lengthy spiel? A friend of mine in St. Paul has found the perfect solution.

As soon as he has identified an unsolicited call, he gently lays the receiver down on the kitchen counter and goes on with his life. It's usually several minutes before the caller realizes that his best performance has been wasted on some nearby kitchen appliances.

While the caller rambles on incessantly about credit card insurance, hail-proof siding or a powerful vacuum cleaner with twin turbine motors, it's possible to strip the sheets off all the beds or put together tomorrow's casserole.

By picking up the phone at various intervals and inserting a random "Let me think" or "I don't know" into the prepared monologue, the call can be extended even further.

If we are creative enough as a nation to send men to the moon or to develop wonders like microwave popcorn, we should be able to discourage unwanted phone calls.

Ultimate Reality Dolls

Move over, Barbie, and make way for the Happy to Be Me doll.

Displeased with Barbie's measurements, which translate to 36-18-33 in real life, an independent doll designer, Cathy Meredig, came up with a more realistic companion for pint-sized consumers.

The Happy to Be Me doll, which should be on toy shelves next month, has more down-to-earth measurements: 36-27-38.

Apparently, too many little girls and boys expect to look like Barbie or to marry a life-sized version when they grow up. The new doll should curb some of that bitter disappointment.

I would suggest that the designer go back to the drawing boards. The new doll still has an hourglass figure, and according to my latest survey of customers at the grocery store and the last time I saw myself in a full-length mirror, there aren't too many of those around.

It's not that my figure doesn't resemble a timepiece. Rather than an hourglass, it could better be compared to a round sundial or London's Big Ben.

If dolls' figures continue to evolve, reflecting this new trend toward realism, we might soon see the What-You-See-is-What-You-Get doll.

With measurements of 36-34-38, the WYSIWYG doll would come complete with cellulite deposits on her thighs, oven burns on her forearms, unusual bulges on the inside of her knees and an interesting array of varicose veins. A few gray hairs and crows' feet might be thrown in for additional effects.

While Meredig wanted a doll that "doesn't look like something out of an old man's dream", the WYSIWYG doll would look like someone the same old man might see across the breakfast table.

The WYSIWYG doll wouldn't mean spending the equivalent of the national debt on a huge wardrobe and accessories. She would be content with elastic-waist slacks, one-size-fits-all T-shirts, a faded bathrobe and sweat pants.

Although doll manufacturers have a long way to go before they duplicate the image of the real American woman, designer Cathy Meredig should be acknowledged for this first great step into toy store realism.

Hip-hip-hurrah!

Christmas Resolutions

Next Christmas will be different.

Inspired by a couple in northern Minnesota, who in turn were undoubtedly inspired by a past president calling for a thousand points of light, my first New Year's resolution deals with our outdoor lighting plans for next Christmas.

The couple's home was featured in a recent newspaper story. Complete with 10,000 lights, half a dozen Santas and countless nativity scenes, the well-lit house and yard stood as a tribute to Thomas Edison. It wouldn't have been surprising if pilots flying overhead mistook the place for Las Vegas.

Our lights didn't go up this year. Actually, it's hard to be creative year after year with a string of 12 outdoor lights. One year we ran the lights up one side of the front door. During another year, the lights were left in a tangled clump on a shrub in our front yard.

While some people take as long as two months to decorate their house for the holidays, our lights require at the most five minutes, two nails and one long extension cord.

After giving the outside lighting project some thought this year, we decided that rather than putting people in the holiday mood, our meager lights were actually depressing them.

Yes, next year will be different. We've decided to add another string of lights, bringing the total count up to 24.

People will no doubt turn out in droves to see our front door, which will be completely surrounded by lights. The police department will have to add extra men for crowd control in our front yard.

My second resolution has to do with homemade gifts. For more years than I care to remember, gift exchanges have been a constant source of embarrassment. While I hand out gifts bought in stores, everyone else is handing out spectacular gifts they've either knitted, crocheted, sewn, hammered together or embroidered.

I have nothing but admiration for people who can transform bleach bottles into delicate treetop angels, or who can make decorative dolls out of old sweat socks.

Next year I'll make some gifts, even if does mean developing eye-hand coordination. Since the great Sweater Fiasco of 1967, I've been reluctant to try anything else.

That year I decided to knit my husband a sweater for Christmas. I didn't let the fact that I had never knitted before or the fact that I had received a D in "Home Making Skills" deter me.

I tried not to think about the generations of women in my family who had been born with non-functioning thumbs.

Equipped with two menacing needles, a booklet entitled "How to Knit" and enough grey yarn to circle the globe at least three times, I began knitting and purling. Unfortunately, the finished product left much to be desired.

The waistband of the sweater came only halfway down my husband's chest and the sleeves nearly reached the floor. Soaking the sweater in water and trying to stretch it out only made matters worse.

If my husband resembled an ape or a Neanderthal man, there wouldn't have been a problem

This next Christmas, I'll try knitting again. I only hope that each recipient has a good sense of humor.

New Treatment for Snake Bites

Either because of my religious upbringing or watching too many Indiana Jones movies, I have a real fear of snakes.

The logical part of my brain knows that most poisonous snakes live in moss-covered swamps or deserts. In addition, no poisonous snake in its right mind would choose to live in a state with six months of winter.

However, since that part of my brain isn't user-friendly, I know for certain that somewhere nearby there's a snake with my name written on it.

My fear extends to all snakes. It would be just my luck to cross paths with a garter snake with a bad attitude.

I've watched enough westerns to know that a snake-bite can result in hallucinating, uncontrollable sweating and anticipating pain beyond imagination. It's not unlike waiting in the dentist's office for root canal work or going over tax records with an accountant.

Even worse than receiving a snakebite would be taking care of a snakebite on another person.

I've seen John Wayne grimace plenty of times as he would make a crisscross incision with his knife and then suck out the venom. I could tell by the look on his face that he would much rather be eating brussels sprouts or facing a firing squad - two of my other greatest fears.

The snakebite victim, either a young schoolmarm disenchanted with life out East or a grizzly old prospector with a heart of gold, would cry out in anguish and writhe on the ground. John Wayne would pull out his knife and say, "We-e-ll now, pilgrim, this is going to hurt some."

In light of the fact that disjointing a chicken or slicing raw liver makes me squeamish, cutting into a person's leg would be even more unsettling.

I was relieved last week to read that sucking venom out of a snakebite is no longer an acceptable practice. Evidently, the person doing the cutting and sucking might land up with more poison in his system than the person bitten.

Dr. Myron Harmaty, who practices medicine in Gastonia, NC, has issued this advice and, because of his familiarity with snakebites, I'm also convinced that there's no way on earth I would live in North Carolina.

He advises us to have the victim lie down to slow the spread of the venom. We are then told to get help from the nearest emergency facility.

Dialing 911 sounds much better than giving a love-bite to someone's leg.

Talking Sports

For all I knew, my friend was speaking a foreign language.

"How about those Twins?" she asked. Her voice rose to a feverish pitch. "The Twins fjeight kshe eithtud theis e tien fie st ir rlds!"

I'm certain those were her exact words.

For a few confused seconds, I knew what life was like near the Tower of Babel. My recovery was quick once I realized I was talking to a sports fan, rather than a foreign visitor who had taken the wrong turn off the interstate.

"Could you repeat what you just said?" I asked.

"I said - Scott Erickson (blah blah blah) ERA (blah blah blah) the magic number is one (blah blah blah) heading for the American League playoffs."

"That's just great!" I told her. "You're talking about baseball, aren't you?"

She looked at me in disbelief and then emitted a long sigh. I've grown used to those looks and disappointed sighs for most of my life. It's not easy living in a spectator world when you don't know the difference between a outfielder's glove and a hockey puck.

H.L. Mencken once wrote, "I hate all sports as rabidly as a person who likes sports hates common sense." Unlike H.L., I don't hate sports - I simply don't understand them.

I can't understand why football officials lug chains along the sidelines while they never seem to need them on the field.

I can't understand why two boxers will hit each other until one becomes unconscious.

It's a mystery why people will stand in long lines to buy basketball tickets so they can sit for hours on hard, wooden bleachers.

I can't understand why hockey fans cheer when their favorite players are sent to the penalty box.

I don't understand the customs surrounding a homer hanky.

Perhaps the lack of understanding goes back to my childhood, when boys played sports and girls didn't. Instead, half of the school was relegated to the sidelines. My young friend, the baseball fan, can't imagine those days. She was born 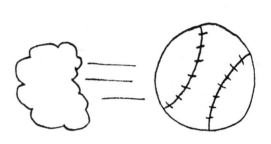 long after someone finally noticed the inequalities of dividing students into two distinct groups - male athletes and pom-pom girls.

It's nice to know I'm not alone. Much attention has been given recently to the mayor of Minneapolis, who apparently knows even less than I do about sports.

Yogi Berra (the ball player, not the cartoon character) must have had people like hizzoner and the rest of us in mind when he observed, "If people don't want to come out to the park, nobody's going to stop them."

Scratch-and-Sniff Pantyhose

Pantyhose have become much more complicated. They're not just for looks anymore.

Thanks to a Japanese firm, American women will soon be able to slip on a pair and have their legs moisturized at the same time.

The new "scratch and sniff" pantyhose will also ward off bothersome insects like mosquitoes and flies.

Laced with tiny microcapsules, the stockings release moisturizing agents or insect repellents as the user moves around. The Osaka firm, Kanebo Ltd., is also marketing pantyhose with seaweed and vitamin C infusions in a London department store.

Although the effects of seaweed on skin are questionable, there must be a good reason why it became an active ingredient in hosiery. Some designer was undoubtedly told to come up with fishnet stockings and something was lost in the translation.

In addition to worrying about wrenching our backs and falling over backwards as we put on the clumsy undergarments, we now have to worry about selecting the correct "scent du jour". Before hurdling feet first into a pair of pantyhose, women will have to predict whether their biggest problem of the day will be an insurmountable attack by mosquitoes or a serious case of dry skin.

If the scented stockings catch on with American consumers, it's conceivable that other scents will appear on the scene. Like pantyhose, the possibilities stretch the imagination.

The woman who chooses to make a fashion statement while she's hunting or fishing might opt for stockings laced with deer scent or "eau de bacon rinds."

In light of the recent controversy over sexual harassment in the workplace, a woman might choose to wear stockings chemically treated with Mace. The stockings would announce to her male co-workers in no uncertain terms, "Look, but don't touch."

Working women suffering from chronic sleep deprivation might choose to wear stockings laced with caffeine. By simply stretching their legs they would receive satisfying jolts to their central nervous systems.

Persistent headaches could be overcome by wearing stockings treated with tiny microcapsules of aspirin.

"Gosh, I have an awful headache," one woman would tell another. "Do you happen to have an extra pair of pantyhose?"

The possibilities are limitless. Pantyhose could soon supply women with their minimum daily requirements of vitamins and minerals or they could administer daily dosages of appetite suppressants.

Unless the marketing plans for the new pantyhose hit a snag, getting dressed for the day will be much more complicated.

A Marriage Based on Competition

The winter games have begun.

I'm not referring to the Winter Olympics, but to the fierce spirit of competition that prevails in our home when my husband and I can no longer do our separate activities outdoors. From the time of the first big snowfall until the sighting of the first robin in the spring, we match our wits over every board game and card game imaginable.

Actually, he came into our competitive relationship with an unfair advantage. At the time of our wedding vows - which should have been rewritten, "To love, honor and to be a good sport" - he already knew that a Scrabble Q-tile was worth 10 points and that a house on Boardwalk could be rented for $200.

It wouldn't be totally surprising to hear that his first words were "Clubs are trump." He undoubtedly teethed on a cribbage board.

While he was brought up on the rules of the games, my upbringing was another matter. Most of the activities in my childhood home were food-related. They included one pastime which could have been called "Who ate all the popcorn?" and our own version of hide-and-seek, "Where did Mother hide the cake?"

By the time I graduated from high school, I was still game-illiterate. Although I knew about mathematical theorems and Shakespeare's sonnets, I knew very little about competition with its sweet victories and bitter defeats.

Things have changed since our marriage. Our winter months have become non-stop marathons of bumper pool, two-handed pinochle, Scrabble and Trivial Pursuit. Winning has become everything. With breaks set aside only for work, major surgeries and sleep, most of our days are devoted to outdoing each other.

Rising to a challenge has become a cornerstone of our marriage, and it's the main reason we stay together. Our vocabularies don't include the phrases, "I give up" or "I'll pass."

Although this secret to a successful marriage has caused our children to renounce games completely and has given them a good excuse to retreat to their rooms, it seems to work.

After the dinner dishes were cleared from the table last night, my husband and I played a game of pinochle. After suffering the slings and arrows of defeat, I rallied with another challenge.

"I'll bet you can't wad up that score paper and hit the wastebasket", I taunted him.

The paper wad arched through the kitchen and scored perfectly.

"Lucky throw," I muttered.

Marriages based on love and mutual respect are nice, but a marriage based on competition is definitely more exciting.

Brain Jogging

There's hope for those of us unable to squeeze into a sweatsuit or unwilling to lift weights. The new exercise is called brain jogging, and it's the latest fitness fad in Europe.

The theory behind brain jogging is the same as for exercising any other part of the body - if you don't use it, you lose it.

As you read this, companies are designing cerebral fitness exercises, which may be individual or group activities. One exercise has you study a complex geometrical design and then draw it from memory. Carried one step further, you are told to draw the design in reverse.

Like physical exercises, the designs should carry some sort of warning from a yet-to-be-named Psychologist General: "These exercises should be used only after consulting with your high school math teacher or someone who really cares."

Coaxing a brain into activity after years of exposure to television sit-coms should be done gradually. After selecting a design which was familiar - the American flag - my final drawing resembled a plate of vermicelli with clam sauce.

Knowing my limitations, I'll stick with squares and circles for the time being.

Another brain-jogging counselor suggests memorizing your grocery list and then leaving it at home. Returning from the store with only half of the items is supposed to motivate you to try a little harder.

Serving a casserole with only half of its ingredients is not my idea of positive motivation. Actually, I've been doing this exercise for years.

It's not uncommon for me to go to the grocery store with the intent of buying a few items on a list left behind on the kitchen counter. Instead of buying sugar, flour and butter, I invariably land up with a car trunk full of everything but the items I really needed.

Caution should be exercised by people hoping to jog their brains. If you can't readily recall your address, your age or today's date, you shouldn't try memorizing geometric theorems, the periodic table of chemical elements or a list of U.S. Presidents.

Once you've mastered the basics, it's safe to advance to more strenuous recall exercises, including your social security number, the names of your children and where you parked the car.

Brain jogging does have certain advantages over physical exercise: the brain can't sweat and you don't need special shoes.

If this fad catches on in other parts of the world, it's possible we may hear future Olympic officials cry out, "On your mark, get set...think!"

Emergency Snow Removal Routes

George Orwell would have appreciated our cable television system. As you may recall, citizens in his novel, "1984", were electronically monitored by their government.

While our local cable system may not be keeping track of subversives and nonconformists, it certainly has strong feelings about street parking during snow emergencies.

It's not as though people in our town just arrived here on a banana boat. We've endured countless snowfalls this winter and warning signs about snow removal are posted at two-foot intervals along the curbs.

During any significant snowfall - when we least suspect it - all of the cable television programs are interrupted by a shrill, ear-splintering sound. It's not unlike the shrieking one hears before being trampled by a raging, bull elephant. The television screens go blank, and a stranger to the situation might give serious thought to running to the nearest air raid shelter.

Rather than showing a subtle, printed message across the bottom of the screen and allowing programs to run their natural course, the cable service chooses a more direct approach. The general assumption seems to be that most television viewers are illiterate.

The voice of a woman, sounding like any mother on a bad day, intones, "I'm not going to tell you again. It's snowing outside, so you'd better move those cars off the street. If you fail to do so, you will be grounded for two weeks."

All right. Those may not be her exact words, but

it's clear that the message is more important than any program we may be watching.

Actually, we know when to expect the announcements. It's almost a certainty they will appear when the program we're watching is in its final, dramatic minutes.

"John, I can't marry you. The baby isn't yours. The father is...Blaaah! Please remove your cars from snow emergency routes immediately..."

"You're probably wondering why you've all been called here. We now know the identity of the murderer. He is...Blaaah! Please remove your cars from snow emergency routes immediately..."

By the time the car removal announcements are over, the final production credits are rolling on the screen or a commercial is extolling the virtues of the latest in hemorrhoid preparations. We have to wait until summer reruns to learn the name of the father and the killer. Reading the final scores of the game of the century the next day isn't quite the same as being there.

Yes, Orwell would have loved it.

Puzzling Popcorn

While the rest of us have been losing sleep over the Brazilian rainforests, it appears the people down there have their own priorities.

According to a recent article in Nature magazine, Brazilian scientists have been analyzing popcorn. Their findings may not make the Nobel Prize committee sit up and take notice, but they do make interesting reading.

For instance, have you ever given serious thought to the thermal diffusivity of the pericarp, the kernel's tough outer layer? How many times have you and fellow co-workers argued about the starch content of popcorn's endosperm?

I thought so. While most of the world worries needlessly about the greenhouse effect and the possibility of walking outdoors and being flash fried, some of the greatest minds in Brazil are trying to figure out why some popcorn kernels pop and others don't.

What fools we have been.

While the article is quick to explain that the pericarp in popcorn has more densely packed fiber than ordinary corn, it fails to answer several other questions.

For example, why isn't there special dental insurance for people who like to chew unpopped kernels from the bottom of the bowl? As someone who has spent the equivalent of the national debt on replacement crowns and caps, I also wonder why the destructive kernels are called "old maids" rather than "bachelors". Is this some sort of agricultural, sexist movement? If this truly is a case of food harassment, should we complain to Gloria Steinem or the FDA?

Microwave popcorns come in a wide variety of flavors - butter, imitation butter, theatre butter, caramel, cheese and white cheddar. When will the pop-corn industry finally come up with a flavor we could really get our teeth into - chocolate-covered popcorn?

If they can't rise to the challenge, how about a pop-corn that can be chewed by people too young or too old to have teeth?

My main complaint about popcorn deals with those troublesome little seed coverings that cling to the roof of my mouth. Resembling small suction cups, they are impervious to tongue gymnastics or finger proddings. Either that or the seed coverings become wedged between my teeth, giving me the sense of having three inches of teeth in a two-inch space.

One of my fondest childhood memories was watching my uncle's patch of field corn and imagining what would happen if the scorch-ing sun would cause the corn to pop. Wouldn't it be glorious if popcorn scientists could develop a variety that would pop in fields on holidays like the Fourth of July? Fireworks explosions would pale by compari-son.

On the other hand, if the Brazilian scientists stay focused on popcorn rather than rainforests, the days will become hot enough, anyway.

The Old Man

The words still strike terror in my heart - "mean old neighbor".

The mean old neighbor from my childhood was partly Freddy Kruger, the nightmare on Elm Street, and partly CIA. The kids in my neighborhood knew better than to set a foot on his property. He had special, built-in radar which could detect the slightest movement within a fifty-food radius of his carefully manicured lawn.

We would no more step on his property than we would step on a sidewalk crack and break our mother's back. He regarded dandelions and creeping jenny as personal enemies, and he was forever digging them up by their roots with a curious vengeance.

If a bicycle tire would brush the edge of his front sidewalk or if an unsuspecting youngster would attempt a shortcut through his yard, he would fly out his front door with his white hair in disarray, shouting a litany of curses.

"Get off my grass! I know who you are and I'll be talking to your parents!"

He was called "the mean old neighbor" by the younger kids on our block. The sophisticates among us called him simply "the Old Man". We didn't know much about his personal life. We did know for sure that he was too old to have ever had children.

He no doubt would have agreed with writer Helen Castle, who once wrote, "Give the neighbor kids an inch and they'll take a yard."

The Old Man's edicts and threats created a challenge for us during the summers.

"Betcha don't dare cross through the Old Man's backyard!"

For extra emphasis, we would add, "Double betcha!"

Last week, when we least expected it, my husband and I became the "mean old neighbors".

Children come and go in neighborhoods and at present ours resembles a playground. Except for the memorable scene when Dorothy encounters the Munchkins in "The Wizard of Oz", I've never seen so many short people gathered in one place.

It's not as though we don't like children. We had four of them to prove the point. But when I looked out the kitchen window and saw a small girl of about four bending one of our seedling trees down to the ground, I thought I should inform my husband.

He stuck his head (complete with white hair in disarray) out the back door and gently asked, "Say, little girl, would you please stop bending that tree?"

It was definitely a polite request. As I watched from the kitchen window, the little girl's reaction was quite a surprise. I suddenly understood the expression, "killing someone with kindness". Without a word, she darted from the yard.

In retrospect, the tree wasn't really that big a deal. The most disconcerting thing was realizing we had become the nightmares of our childhood.

Blood Tests - Just Say "No"

A recent Mayo Clinic report proves something I've suspected for a long time - we're being sucked dry with blood tests.

Those weren't their exact words, but the researchers found out that 45 times more blood was taken than was needed for basic tests. They also found that blood was withdrawn more times than was necessary.

This may not be exciting news for people with two-inch thick veins running up and down their arms, doing everything but saying, "Take me! Take me!" But for those of us with no veins, the report might keep us from becoming human pincushions.

Although I know it's medically impossible not to have veins and still be alive, my veins are cowardly things. At the first sign of a needle, they run for cover. They have been responsible for making several nurses and lab technicians consider other professions.

My medical abnormality was discovered while I was still young. My husband-to-be and I had gone to the doctor's office for the mandatory, pre-marriage blood tests. I realize now that the tests weren't conducted to find out if our blood was wet and red, but rather it was a preliminary stress test. Any couple making it through the ordeal was ready for the real test: marriage.

I can remember the nurse assigned to us vividly. She stood about five foot-three and her commanding girth was enveloped in a nubby red cardigan.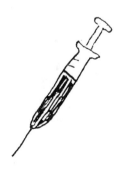

"Roll up your sleeves," she barked at my husband-to-be. Once he had done so, she suddenly became soft and mellow.

"Nice veins," she purred. It was

obvious he had her instant respect because of his particular circulatory system. In the shortest time imaginable, they had exchanged pleasantries about the college we were attending and the weather, and she had siphoned the necessary one gallon of blood. With a fresh needle in hand, she turned to me.

"You're next, honey."

In what they interpreted to be a call to retreat, my veins suddenly went AWOL.

"I don't have any veins, " I apologized.

"Why don't we just have a looksee, anyway?" she laughed. "I always find a good one."

The next few minutes seemed like hours. She poked and prodded until beads of sweat formed on her forehead. She wrapped various parts of my arms in elastic straps until I was sure the circulation had been cut off completely. I wondered how my wedding dress would look with one empty sleeve dangling loosely at my side.

She tried one arm and then the other. At times she would grin determinedly and jab the needle into my arm, reminding me of a person trying to get the last dab of peanut butter out of a jar. At one point, she straddled the arms of my chair like a jockey riding down the final stretch.

"Doctor's gone already. You'll have to come back tomorrow."

For the next 20 hours I had second thoughts about marriage. Suppose we should get married and have children. Would it mean more blood tests? Would our marriage amount to anything but a long series of exasperated nurses? Is this what marriage is all about?

We were married and it was a lovely ceremony. The long sleeves of my gown covered up the black-and-blue arms nicely. Everyone said they had never seen a bride sob so much at her own wedding.

Shorthand English

Certain words are disappearing from our language at an alarming rate. Before you panic and look up the number of the grammar police, you should know that two of these words are "from" and "the".

It's possible that in the very near future we will hear and read sentences like, "He received gift doctor", instead of, "He received the gift from the doctor."

Given the way these two words have been dropped from everyday usage, it's also very possible that our entire language will eventually be reduced to grunts, shrugs and a few nouns.

A recent headline was my first clue that something was amiss: "Clintons see Chelsea graduate eighth grade." The second red flare went up when a nurse announced, "Doctor will see you now", and later, "You don't want to be weighed? Let's see what doctor says."

After the headline incident I became more aware of people saying so-and-so graduated college or someone else had graduated high school. Now, I'm not sure what they've been taught in those schools, but I do know that if I would have said "I graduated college" back in 1965, my diploma would have been recalled for possible defects.

I can understand saying "I graduated college" when the school in question was found through a matchbook cover, but most colleges expect more from their graduates.

The doctor thing is another matter. I can't remember ever hearing "Lawyer will see you now" or "Teacher sent home a note".

Is it possible that referring to a doctor as "doctor" rather than "the doctor" is simply a sign of respect? I doubt it. After all, people still refer to "the President".

Even the pope receives a "the". It doesn't sound right to say "He had an audience with pope".

If this trend continues, with prepositions and articles discarded like so many peanut shells, we're all going to land up sounding like movie extras form "1,000,000 B.C."

"I go...work now. Give me kiss."

"Good. When come home ... work, buy groceries ... store."

As our language becomes more and more abbreviated, we will no longer give our children long, cumbersome names like Christopher and Jennifer. Instead, they will be named Zonk and Grog.

No longer content as a country with fast-food restaurants, drive-through funerals and complete meals that can be microwaved in two minutes, we are clearly headed for another drastic change: fast language.

The Birds and Bees

As a child, I never really understood what the birds and bees had to do with having children.

My parents were no help. Whenever I asked my father about my origin, he would ramble on about pollen and other unrelated topics. Eventually, he would say, "Go ask your mother." She, in turn, would send me to my father.

A path was literally worn into the carpet as I rushed from one parent to the other, seeking the sweet mysteries of birth. For the longest time, I seriously wondered how two such ill-informed people ever managed to have children.

In an attempt at self-education, I began watching birds and bees in earnest. The only conclusions I was able to draw were that birds flit and bees sting.

If only I had been born 50 years later and my parents would have had a computer. The mysteries of birth have now been reduced to floppy disks for IBM personal computers in a program called BABY.

As a result, parents who are reluctant to have a one-on-one about such touchy matters, or are too old to remember, only have to say, "Go ask the computer."

In all likelihood, the computer will be totally candid and won't say, "Go ask the refrigerator."

By the time my husband and I became parents, the birds and the bees were out and Dr. Spock and honesty were in. We knew that when our children would ask we would answer.

When our oldest daughter was three she asked, in between bites of a peanut butter sandwich, "Where do babies come from?" In spite of the books and the child psychology classes, I was caught off guard. I had hoped she would wait to ask until she was a little older. Perhaps 20 or even 22.

Drawing upon a creative, well thought out analogy, I explained the process with the description of a garden and seeds. By the time I had included Mommy the garden, Daddy the gardener, the seeds growing into babies and others in a cast of colorful characters, she had finished her sandwich and I was left alone, babbling to myself in an empty kitchen.

As a sex educator I considered myself to be a dismal failure. There had been no follow-up questions. I didn't even have a chance to fall back on the birds and bees story.

About a week later, she was back at the kitchen table, shaping her macaroni and cheese into a meaningful pattern with her fork.

"That must be a nice store," she said.

"What store? What are you talking about?"

"You know. The store where they sell baby seeds."

She didn't ask any more questions and I didn't bother to explain. Her version was better than mine, anyway.

By the time her younger sister began asking about babies, I was much more direct and to the point.

"Go ask you sister."

The Wedding - Part One

Like most mothers, I want the best for my daughters.

While they are growing up, I hoped their futures would include exciting careers, thin thighs, homes with white picket fences and children with all the right genes.

I even dreamed about handsome Sir Galahads who would sweep them off their feet and carry them to faraway kingdoms after fairy tale weddings. The Galahad fantasies were forms of escapism. They were most prevalent during unsuccessful attempts at toilet training and during tearful, sleepless nights caused by teething. Actually, the girls teethed and I cried.

Our oldest daughter has met her Galahad, and next June they will begin a life together in Chicago, apparently the next best thing to a faraway kingdom by today's standards.

I had often imagined what would happen when a daughter would announce her engagement. Although the details varied from time to time, I knew that I would be much, much older.

I would be sitting next to the fireplace, knitting a large afghan, and my husband would be sitting near me, reading the newspaper. The front door would fly open, and one of our daughters, blushing and giddy, would rush to our sides, flashing a ring with a large stone.

"Oh, dear parents," she would announce breathlessly, "Harold and I are engaged! Isn't that wonderful?"

"Darling, we're so happy for you," I would respond. I would be very composed, because when you are much older, composure comes with the territory.

In reality, the announcement came over the phone lines from Chicago.

"We're engaged!" our daughter shouted.

Caught off guard, I asked, "Who is this?"

"This is Kris. Your oldest daughter."

"Great news, Kris." Then I paused. "Does this mean I'll have to wear one of those frumpy, beige mother-to-the-bride dresses done in lace and chiffon?"

"Wear anything you want, Mom." As she spoke, I settled on denim.

My next reaction to the announcement was totally unexpected. I immediately regretted ever laughing at mother-in-law jokes in the past. It suddenly dawned on me that in a few months I would become one of "them".

The next day, I bought two of those magazines devoted to brides and wedding plans. After flipping through the pages and reading article titles ("10,000 Things to Do Before the Big Day" and "The High Costs of Getting Married"), I realize that fairy tale weddings occur only in stories.

I do have one advantage. I was a high school prom advisor for 10 years and it appears that proms and weddings have much in common. All the women wear dresses that can only be worn once and both events feature imprinted napkins, flowers and crepe paper streamers.

The "10,000 Things to Do Before the Big Day" article does bother me, though. No mention is given to cleaning the house or losing weight. The thought of doing 10,000 things during the next seven months is mind-boggling.

Fantasies aren't supposed to turn out this way.

The Wedding - Part Two

Perhaps it starts with a reception following high school graduation. Or maybe it's a wedding.

Regardless of the cause, it's a multi-million dollar business in this country, and it goes by a simple name - "fixing up the house".

In an attempt to convince relatives and friends from far away that we live better than we actually do - or that our numbers came up in the lottery - we rip out carpeting, knock out walls and slather every surface with paint.

When our guests arrive, we say things like, "What a lovely surprise! Please forgive the mess". We don't want them to suspect that we were still tearing price tags off the furniture when the doorbell rang.

We don't want them to know that our families have been eating their meals behind the basement furnace so the new carpeting upstairs won't have any spill marks. We don't let on that the white brocade sofa has never come into direct contact with a human body.

The most amazing thing about "fixing up the house" is that not once does someone say, "Let's go all out for graduation this year. Let's completely redo the house and ring up enormous debts, which will undoubtedly send us to the poorhouse."

Instead, the entire process usually begins with the purchase of something simple - something small "to spruce up the place a bit." It could be a lamp, or a bargain throw rug to throw over the spot where the dog got sick from eating too much pizza.

From this point on, things begin to snowball. Renovation becomes an obsession.

The lamp looks great, but once it's placed on the end table by the sofa, the furniture looks like something dragged in by the cat. There's no doubt about it - the furniture has to go.

Before we know it, we've received estimates from carpenters, painters and wallpaper hangers. In a fixing-up frenzy, we roam from room to room, condemning to the landfill every object in sight - from bathroom fixtures to picture frames.

If there's enough time before the reception, we may spread out to the garage and yard. As we try to make everything look just right for a two-hour event, we even look with scrutiny at our family members.

"You! Yes, you, with the red hair. I hate to tell you this, but you're clashing with the kitchen."

Regardless of what food is served at the reception, it's a sure thing that every morsel will be consumed. Either the family members haven't had a meal prepared for them in weeks or they've been too busy - "fixing up".

The Wedding - Part Three

"If the wedding is to be a large formal one, expert help should be arranged... Otherwise father may be irritated, mother jittery, the bride in tears and the groom cross."

Emily Post's Etiquette
1960

Color me jittery.

A few months ago, when I realized our daughter's upcoming wedding would be more complicated than a family dinner, I decided to let my fingers do the walking and find a wedding consultant in the telephone directory.

Unfortunately, there were no listings between Waterproofing Contractors and Weed Control Service. Although both of those professions undoubtedly serve definite needs in our society, it became clear that unless the wedding was going to be held in a damp basement with bouquets of dandelions, a wedding consultant wasn't in my future.

Since the wedding date was set, my life has become a series of madcap adventures. A cake lady in Iowa wants to show me photographs of her favorite creations. ("A chocolate cake? That's no problem. I'm also doing some very exciting things with carrot cake.")

Because of the wedding plans, I have met some extremely interesting people, who apparently will stop at nothing to make our daughter's big day one we and our banker will never forget.

One morning last week, I visited with the photographer. While I was interested in basic things, like cost and whether the photos will make me look 20 years younger, he seemed more interested in creating the right atmosphere for our little discussion.

The quality of his work was never called into question. His main problem was working with the wrong mother.

After a formal exchange of names and handshakes, I was ushered into a viewing room and led to a beautifully upholstered Queen Anne's chair. As the door was closed, the lights dimmed and what can be best described as "prom music" ("This night is ours forever...la-la-la") began playing from several unseen speakers.

A rapid succession of blown-up wedding photos began appearing on the wall, and my senses soon became short-circuited.

"This is crazy. This can't be happening," I thought. For some unknown reason, I began giggling uncontrollably.

Perhaps I had seen one too many sample books of wedding invitations or too many decisions about cake recipes had put me over the edge. As I sat there in the dark, with elevator music in the background and eight-foot brides smiling at me from the wall, I fulfilled Emily Post's prophecy - I became jittery.

Still jittery, I called the bride-to-be in Chicago when I returned home.

"Mom, what you're feeling is only normal. You have to get used to being a mother-in-law."

I didn't have the heart to tell her, that after 26 years, I'm still getting used to being a mother.

The Wedding - Part Four

Planning a wedding really takes the cake.

Last Saturday my daughter and I met with a cake lady for almost two hours. In the same time that it took to draw up the Marshall Plan for postwar Europe or the time that it takes to deliver a set of quintuplets, we talked cakes.

My original intent had been to meet briefly with the woman, tell her how many people we were hoping to feed and figure out some sort of payment plan.

After all, a cake is a cake. Although I'm suspicious of any cake recipe calling for tomato soup, sauerkraut or zucchini, I've never met a cake I didn't like.

However, for this particular woman, cake is more than a staple commodity like bread or vegetables. Cake is more than a happy mixture of flour, sugar, eggs and milk - cake is a calling. I should have known this when I first spotted her lawn ornament, shaped like a gigantic wedding cake, or the teetering towers of cake design magazines on her kitchen table.

As we cakewalked through the magazines, it became clear that a wedding cake is more than a fancy confection with a plastic bride and groom on the top. Given enough pictures, planning a cake can become an out-of-the-body experience.

Many of the designs were clearly variations of the same theme. A cake with two stairways and one fountain. A cake with three stairways, two fountains and doves. A cake with a flock of doves, the New York skyline and miniature strobe lights.

It was while we were looking at the cakes of a California designer, creations done in shades of purple and green, that I short-circuited. There were simply too many pictures and too many choices.

"Pick any cake," I whispered to my daughter. "They all look the same to me."

A quick exit wasn't in my future. We still had to discuss frosting textures (chantilly lace or basket weave?) and whether the plastic pedestals would be placed between the second and third or first and second tiers of cake.

While the woman and my daughter discussed structure and design, I wondered whether I had done my daughter an disservice by not picking up a degree in architecture.

The rest of the meeting was a blur. As my daughter single-handedly made some apparently wise choices about frosting flavors (almond or vanilla?) and whether the frosting color should be off-white, cream white or ecru, I clung desperately to fragments of reality.

I tried to recall my social security number and the names of my children to no avail. As the three of us sat there, determining which design would make or break this special day of all days, one thing became totally clear.

Picking out a wedding cake definitely isn't a piece of cake.

The Wedding - Part Five

When our daughter called to tell us about her engagement, my first response was, "I won't have to wear one of those dreadful dresses done in beige chiffon and lace, will I?"

"Wear whatever you want," she sighed. Like most daughters, she has perfected the art of heavy sighing, rolling her eyes upward and saying "Oh, mother!"

With the wedding day only a month away, my phone has been ringing off the hook because people want to ask the same two questions: "What are the wedding colors?" and "What are you going to wear?"

Before this wedding business came up, my phone conversations were much more interesting. People who knew me well would talk about the weather and events in the news. People who didn't know me very well would ask, "How about those Twins?"

Everyone seems obsessed with what I'm going to wear. I doubt whether Princess Di's mother had to put up with the same concerns.

"I'll just throw something on," I tell them. After long periods of silence, I'm usually bombarded with offers to take me shopping and all sorts of unsolicited advice. Clothes aren't that big a deal for me. If I'm dressing casually, I grab something from a drawer. For more dressy events, I grab something off a hanger. Keeping in style ranks right up there with cleaning the oven and opening clogged sink drains.

I just don't get it. As long as I'm covered with anything, people should be happy.

Another factor also must be considered before I jump into a car and shop for a dress. I have no idea what size I'll be in 30 days. There's always an outside chance that a diet may take hold and I'll lose 20 pounds. On the other hand, if there's a pizza coupon in the newspaper during the next few weeks, I might easily gain that amount.

A closeout on Christmas chocolates might also put me over the edge.

I must admit that all of the questions about what I'm going to wear are starting to get to me. In resignation, I recently picked up one of thousands of wedding guides. Rather than simply listing what people are supposed to wear, the guide goes into a rather flowery narrative about the perfect wedding scenario.

"Your handsome groom will proclaim his love in a tuxedo as your tearful mother stands by in her floor or ankle length dress," reads the guide. It's not exactly Pulitzer Prize material, but it must appeal to someone.

As for the tearful part, I think I can handle that. If nothing else, it will detract from what I'm wearing.

The Wedding - Part Six

With our daughter's wedding only two weeks away, I have become obsessed with chicken prices and guest lists.

Although reading newspapers has become a luxury I can't afford these days, one article did manage to catch my eye: "At the Houston Stock Show in March, the grand champion pig brought $70,000; the prize steer, $200,000, and three broiler chickens were sold for $25,000 each."

Actually, I wasn't all that surprised when I read the prices. I strongly suspect that those same chickens are being served at our daughter's wedding dinner.

When we had to decide what to serve at the dinner, our choices sounded like phrases chosen at random from foreign dictionaries: "prime rib marnier", "filet mignon bearnaise", "c'est la vie", "Gesundheit" and "sayonara."

Partly because we chose not to duplicate the federal debt, and partly because it's easier to pronounce, we chose the chicken, which was listed in decidedly smaller print at the bottom of the menu.

In spite of our decision, our caterer has shown a great deal of class. Unlike Martin Short's caterer role in the recently revived "Father of the Bride", not once has he referred to our dinner choice as "cheaper cheeken". He didn't wince when I suggested having a lettuce salad, rather than rare, imported seaweed fluff.

It's to his credit that he treats us like normal people.

At first, we planned on feeding 100 people. Although 20,000 were invited, we were sure that because of the great travel distance and superstition, which would keep some people from traveling on the 13th, few would be able to attend.

One card file was set aside for "yes's" and another was labeled "no's." With two weeks remaining before the wedding vows are exchanged and our debts are tallied, the score is 200-5.

While the "yes" file of response cards is bursting at the seams, the "no" file is gathering cobwebs. The responses are nothing short of surprising.

The uncle, who dropped out of society during Lyndon Johnson's presidency, has decided to re-enter the mainstream at the wedding.

The aunt, who has had numerous by-pass surgeries and most of her vital organs replaced at one time or another, is coming. Although she has been told repeatedly that she will never laugh again, she will join the festivities with her entire medical entourage.

Although a few of the response cards are signed by names I recognize, most of them are strangers. If I didn't know better, I would suspect that the happy couple casually selected names from out-of-state phonebooks.

A conversation with our second daughter helped put weddings into their proper perspective.

"Now, let's get this straight," she observed. "You're inviting 20,000 people, most of whom you've never met and most likely will never meet again, right? You feed them a big dinner and then thank them for coming, right?"

Right. In a way, she has a point.

Throwing a wedding is a complicated process. By the time most parents catch on how to do it, they've run out of daughters.

I only hope that the mystery guests will enjoy the "cheaper cheeken."

The Wedding - Part Seven

Someone - perhaps Yogi Berra - once observed, "It's deja vu all over again."

I thought about those words as the final hours before our daughter's wedding ticked off this week. Although this is the first wedding among our children, I have the eerie sense that I've gone through this before.

Today, in a rare moment of enlightenment, the parallel experience became quite clear - the week before her wedding is very similar to the week before she was born.

Actually, because of the cake, the church, the tears and the unwritten rule that the women have to wear pantyhose, there are many similarities between weddings and funerals. But that's another column.

For the past several days, my phone has been ringing off the hook with callers offering advice and support.

"How are you?" they ask.

"Fine", I reply, even though the wedding plans have turned my kitchen into a nightmare version of an office and the pile of bills on the table are giving me a case of cardiac arrest.

"But", they insist, "how are you REALLY?"

Outside of the fact that the wedding plans have driven me to thoughts of punching complete strangers in their faces, and I'm having sleepless nights worrying about bows falling off church pews, it's hard to convince people that I really am "fine".

For some unknown reason, people are also choosing these final days to tell me horror stories of weddings gone wrong. Either the mother-of-the-bride's half slip fell off as she was walking down the aisle or 200 wedding guests were sent to the hospital with food poisoning.

The stories sound much like the "giving birth" stories told to me before our daughter was born.

Except for the cover stories on tabloids in the grocery store checkout lane ("Mom Gives Birth to Giraffe"), and the stories told to me in hushed tones by other women, I was fine then, too.

Why shouldn't I be fine? I was eating enough to feed a third world country and I was setting a world record for most weight gained during a pregnancy.

"I probably shouldn't tell you this since you're expecting any day," they would begin. Although I would be devouring an entire cheesecake to feed my anxiety, they would then proceed to tell me about the cousin (or aunt, or sister-in-law) who went into labor for 1,000 hours. Or else I would hear about the baby boy who was colicky until his second year in college.

At this point, I need stories about ruined weddings as much as I need a doctor to tell me that I should lose 20 pounds.

Where are the successful wedding stories when I could really use them? I'm fine, really.

The Wedding - Part Eight

Except for the one small glitch, our daughter's wedding was perfect. The bride was beautiful. The groom was handsome. I was nearly impaled by a piece of jewelry.

Perhaps it would have helped if I would have tried on the mother-of-the-bride dress one more time before the wedding. Instead, I found myself standing before a mirror only minutes before the wedding with a neckline normally found in a French bordello.

I had to do some fast thinking. I could either show up at the wedding, looking like an actress from a B-Grade movie or I would have to batten the hatches.

Precious minutes ticked by as I searched without success for a pin or brooch. In desperation, I gave serious thought to using a paper-clip or contact cement. Suddenly, a rather artistic-looking pierced earring caught my eye. I hurriedly pierced the two layers of cloth with the makeshift pin and left for the wedding.

Soonafter, I learned the difference between public pain and private pain. It's one thing to break a leg skiing at Vail, to limp around on crutches and to tell other people about what happened. It's quite another to hold a dress together with an earring with a long, sword-like prong.

The first hug I received at the reception showed me the error of my ways. Much later, I would discover that my sternum resembled a well-worn pincushion.

After a particularly robust bear-hug, one guest stood back and exclaimed, "Don't look so worried! Everything's wonderful!"

I didn't have the heart to tell him that my facial expression was the direct result of a punctured lung, rather than worry over some social faux pas.

Hug after hug, dance after dance, the pierced earring kept a fixed grimace on my face. Although I've often heard that we must suffer before we can enjoy, I've never heard of the two sensations happening at the same time.

It wasn't the pain of an unassisted childbirth, but the pain was exquisite. How I envied the other women at the reception who wore pins with actual clasps! With my pierced chest, I would have given anything - my car, my house, my Perry Como records - for a bandaid.

Toward the end of the evening, one woman gushed over the attractive pin, which kept me from being arrested for indecent exposure.

"How attractive!" she cooed. "Where did you ever find it?"

I was much too modest to tell her that I had another one like it at home.

Driver License Weights

According to a recent article in the *Star Tribune*, my weight is no longer a secret.

I had always assumed that when I mumbled my weight at the time of my driver's license renewal, that information would be classified as "top secret". Even under the threat of sheer torture, officials at the local and state level would refuse to divulge my true weight. Rather than betray a sacred trust, each government employee would be instructed to swallow a capsule of cyanide hidden in a suit lapel.

Before the final effects of the lethal poison would settle in, their last words would be, "I can't tell you what the woman weighs!"

The article went on to explain how the state collects millions of dollars each year selling data from driver's license and motor vehicle records. The information is freely shared with insurance agencies, financial institutions, retail sales outlets, the Selective Service, direct-mail marketers and collection agencies.

So much for the cyanide capsule idea. As a result of the newspaper article, I have devised a clever plan to thwart the retailer who have been sending me catalogs for large-size pantyhose and weight-loss programs.

On my next visit to the driver's license bureau, I will go to great lengths to look small. I will try not to bump into desks and doorframes with my hips as I walk into the office. I will wear black clothing, which most people think of as slenderizing. I will suck in my cheeks, hoping to give my face the appearance of someone with an eating disorder.

When the clerk looks at my old license and asks, "Is your weight the same?", I won't bat an eyelash.

"No," I'll declare, so anyone within a three-block radius of the courthouse will hear, "you must have me confused with someone else. It's 100 pounds."

Even though that number comes closer to the weight of one of my thighs or what I weighed in seventh grade, I will do what I must to protect my privacy.

I will give them my address, my height and my eye color, but why should complete strangers know my weight? Not even my husband and children know that gruesome detail.

My new weight should come as a surprise to many people. Imagine the look on the mailman's face as he delivers petite size catalogs to our front door. Imagine the reactions of houseguests when they see those catalogs displayed on the coffee table in the living room.

Optimism will soar at a new high at certain life insurance companies when they learn of the new, low-risk prospect in Minnesota - a healthy, strapping woman who tips the scales at 100 pounds. It's clear I will live forever and they will never have to pay off my survivors.

When it comes to data privacy - turnabout is fair play.

Dairy Discrimination

Time has a way of healing all wounds. For example, I no longer harbor grudges about the beauty queens from my high school days. I know my classmates chose them over me because they were too young and foolish to know better.

Now that those beauty queens and I have a common denominator - middle-age - it's comforting to know we are all faced with similar problems. We all know the anguish of hormone shut-downs, finding pantyhose that really fit and hoping to retire before we're 80.

I also hope that dairy cows will be half as understanding when they hear about the preferential treatment being given to certain bovine beauties.

According to an Associated Press story, University of Wisconsin dairy geneticist Denny Funk recently told Midwest farmers the main reason why their milk production is falling behind that of California producers.

He reasons that Midwest farmers have a fondness for keeping better-looking cows around, even if they produce less milk.

Even if what the geneticist said is true, I can't imagine a dairy farmer choosing a pretty cow face over bottom-line profit. Unlike my high school classmates, most dairy farmers I know are old enough to know better.

"According to our records, dear, old Nellie, number 33, is down to a pint a day. Maybe we should have her culled from the herd."

"Aw, give her a break, will you? Have you ever noticed how she sashays through the lot, with that cute little wiggle in her hips? And what about those long eyelashes? When she rolls those big brown eyes and flutters those eyelashes, I just can't think straight."

Choosing a certain cow over others just because she's more attractive would also have a de-mooralizing effect on the rest of the herd.

Being a dairy cow can't be easy work. Besides having to contend with four stomachs and an assembly-line production of calves, she has to share her mate with 30 other females. If she fails to comply with the rules and isn't attractive, there's always the possibility of becoming a 99-cent special at a fast-food restaurant.

No cow in her right mind would tolerate the drawbacks of the job if she wasn't treated fairly. In the words of a well-known TV character, the cow would "have a cow".

Favoritism in the workplace - whether it's an office or a stanchion - has the potential of being udderly devastating.

Copies of LIFE WITH A CHANNEL SURFER, NEVER TRUST A SIZE THREE and POTATO CHIPS ARE VEGETABLES may be ordered through:

Carole R. Achterhof
Columnist - Speaker
Rural Route 9061
Spirit Lake, IA 51360

The following order form may be photocopied.

Please RUSH _____copies of LIFE WITH A CHANNEL SURFER, _____ copies of NEVER TRUST A SIZE THREE and/or _____ copies of POTATO CHIPS ARE VEGETABLES at $9.95 each. (An additional $2 for postage and handling is required for orders of 1-4 books.)

Enclosed is my check or money order for $ _____, payable to BARE BONES BOOKS.

NAME_____

STREET_____

CITY _____ STATE _____ ZIP_____